WINDPUMPING HANDBOOK

Sarah Lancashire, Jeff Kenna and Peter Fraenkel

IT Publications 1987

Practical Action Publishing Ltd
25 Albert Street, Rugby, CV21 2SD, Warwickshire, UK
www.practicalactionpublishing.org

© Intermediate Technology Publications 1987

First published 1987 \Digitised 2008
Reprinted in the UK, 2020
Reprinted by Practical Action Publishing
Rugby, Warwickshire UK

ISBN 978 0 946688 34 0

All rights reserved. No part of this publication may be reprinted or reproduced or utilized in any form or by any electronic, mechanical, or other means, now known or hereafter invented, including photocopying and recording, or in any information storage or retrieval system, without the written permission of the publishers.

A catalogue record for this book is available from the British Library.

Since 1974, Practical Action Publishing has published and disseminated books and information in support of international development work throughout the world. Practical Action Publishing is a trading name of Practical Action Publishing Ltd (Company Reg. No. 1159018), the wholly owned publishing company of Practical Action. Practical Action Publishing trades only in support of its parent charity objectives and any profits are covenanted back to Practical Action (Charity Reg. No. 247257, Group VAT Registration No. 880 9924 76).

Preface

Windpumping is an established technology, with over one million windpumps in use worldwide. A windpump needs no fuel, little maintenance and it usually lasts 20 years or more. Designs exist which are suitable for small-scale local manufacture. The aim of this handbook is to help potential users and decision makers take advantage of the benefits that windpumps can offer.

This handbook was first written for a windpump familiarisation seminar held in Nairobi in November 1986. The seminar was organised and presented by I.T. Power, hosted by the Ministry of Water Development of Kenya and funded by the Overseas Development Administration.

Contents

Page

Preface
List of Figures
List of Tables

1. **INTRODUCTION** — 1

    1.1 Purpose of this Handbook — 1
    1.2 Windpump technology is time-proven — 2
         A brief history of windpumps
         Past experience with windpump designs
    1.3 The wind energy resource — 5
         The effect of wind speed
         The effect of air density
         Energy availability
         How to find the amount of energy available from the wind
    1.4 Choices of energy resource — 9

2. **WINDPUMP DESIGN - STATE OF THE ART** — 11

    2.1 Principles of wind energy conversion: Lift and drag — 11
    2.2 Rotor design — 13
         Pitch
         Solidity
         Tip-speed ratio
         Performance coefficient
         Torque
    2.3 Pump types — 17
    2.4 Transmissions, tails and towers — 21
         Transmissions
         Tails
         Towers
    2.5 The feasibility of local manufacture — 23
         The advantages
         Criteria for success
         Types of design suitable for local manufacture

3. **SITE EVALUATION** — 25

    3.1 Assessing the wind regime — 25
         The wind regime parameters needed
         Wind measurement
         Quality of wind data
         Measurement options
         Choosing the windpump site

3.2	Assessing the water requirement	35
	Water for domestic use and animals	
	Water for irrigation	
	How to find the pumping head	
	Borehole yield	
3.3	Sizing the windpump	44
	Volume-head product	
	Sizing the rotor	
	Sizing nomogram	
	Sizing the pump	
3.4	Storage requirement	49

4. IS A WINDPUMP THE BEST OPTION? 52

4.1	The decision route	52
4.2	What are the alternatives?	54
	Diesel engines	
	Solar pumps	
	Handpumps	
	Animal pumps	
4.3	Social and institutional factors	58
	Practical factors	
	Social factors	
	Institutional factors	
4.4	Costing the options	61
	Economic or financial assessment	
	Life-cycle costing	
	Example financial assessment	

5. PROCUREMENT, INSTALLATION AND OPERATION 69

5.1	Specifying and procuring	69
5.2	Installation	70
	The site	
	Receipt of windpump	
	Borehole construction	
	Windpump foundations	
	Erecting the tower and assembling the rotor	
	Building the storage tank and delivery pipe	
	Fences	
5.3	Maintenance and repair	74
5.4	Safety	75

BIBLIOGRAPHY	76
GLOSSARY	77

List of Figures

		Page
1.	Horizontal-axis, multi-bladed windmill	2
2.	Three-bladed horizontal-axis windmill	3
3.	Schematic diagram of Savonius rotor	3
4.	Panamones	4
5.	Schematic diagram of Darrieus wind turbine	4
6.	Global annual average wind speeds	6
7.	Approximate shaft energy output of a windpump rotor for various wind speeds	8
8.	A boat propelled by the drag force of the wind	11
9.	A boat propelled by the lift force of the wind	11
10.	The relative sizes of lift and drag forces for blunt and streamlined objects	12
11.	Generation of lift by an aerofoil	12
12.	Schematic diagram showing angle of pitch of a rotor blade	13
13.	Change of blade pitch with radius	13
14.	Typical torque versus tip-speed ratio and performance coefficient versus tip-speed ratio curves for rotors of varying solidity	16
15.	Schematic diagram to illustrate the effect on windpump operation time of the high starting torque	17
16.	Schematic diagram of a reciprocating positive displacement pump (piston pump)	19
17.	Schematic diagram of a rotary positive displacement pump (progressive cavity or 'Mono' pump)	19
18.	Typical head, flow and efficiency curves for positive displacement pumps	20
19.	Schematic diagrams showing the furling action of a wind rotor in strong winds (bird's eye views)	22

20. Schematic diagram showing approximate wind acceleration factors over a hill — 27

21. Sea breezes — 28

22. Flow chart outlining the steps to be taken when processing wind data — 31

23. Area of turbulence around a building — 33

24. Area of turbulence around trees — 33

25. Effect of ground friction on wind profile — 34

26. Wind profile changes over trees, etc. — 34

27. Schematic diagram to show selection of tower heights to achieve even wind speeds across the whole rotor — 35

28. Soil moisture quantities — 38

29. Rate of crop growth as a function of soil moisture content — 39

30. Schematic diagram showing pumping head — 41

31. Flow chart outlining the steps necessary to size a windpump — 45

32. Windpump rotor sizing nomogram — 47

33. Typical charts for pump sizing by head and average wind speed — 49

34. Cost of water storage depends on the volume — 50

35. Steps required to choose the most appropriate water pumping technology — 53

36. Typical fuel consumptions for small diesel engines — 55

37. Number of handpumps required as a function of water requirement — 57

38. Number of oxen required as a function of water requirement — 58

39. Flow chart showing the steps to be taken in financial assessment — 65

v

List of Tables

		Page
1.	Altitude correction factors for air density	7
2.	Countries known to be manufacturing windpumps in 1986	10
3.	Coefficients for the effect on wind speed of different ground roughnesses	26
4.	Daily water requirement of farm animals	36
5.	Population increase for various annual growth rates	37
6.	Typical irrigation water requirements for Bangladesh and Thailand	40
7.	Headloss in metres per 100m of pipe length for various flow rates and diameters.	43
8.	Discount factors for various discount rates and numbers of years (zero inflation)	63
9.	Discount factors for recurrent costs which have to be paid annually over a number of years, for various discount rates (zero inflation)	63
10.	Advantages and disadvantages of various construction methods for storage tanks	73

CHAPTER 1: INTRODUCTION

1.1 Purpose of this Handbook

Water for people, animals and crop irrigation is an essential need in every country. Frequently this water has to be pumped from the ground; the pumping requires energy. In rural areas this energy has traditionally been provided by people operating hand pumps or animal pumps. Where mechanized power is available it is most commonly an internal combustion engine burning petrol or diesel oil. Recently there has been a growing interest in the new technology of solar-powered water pumps and a revival of interest in windpumps.

There are many good windpump designs, both traditional and modern lighter weight ones, currently available. These machines have high performance and good reliability. The purpose of this Handbook is to provide decision-makers and potential users of windpumps with the basic information on present-day:

- windpump technology
- economics
- sizing to meet domestic or irrigation demand
- procurement
- installation
- maintenance.

It has been assumed throughout the Handbook that the reader is familiar with the basic concepts and units of energy, power, flow, density, etc. A comprehensive bibliography is appended for those readers who wish to study windpumps in greater depth.

1.2 Windpump technology is time-proven

A brief history of windpumps

The ancient Egyptians used wind power 5000 years ago to propel boats. It is uncertain when wind power was first used on land to power rotating machinery but it is estimated to be about 2000 years ago. Historical records show that windmills definitely existed in 200 BC in the area now known as eastern Iran and western Afghanistan. This area receives constant winds from the steppes of Central Asia during and after harvest time each year, called the "Wind of a Hundred Days". The Chinese have used windmills for low lift paddy irrigation for many centuries.

About 1000 years ago horizontal-axis sail windmills were being used around the Mediterranean. By the 12th century windmills had reached northern Europe. They became an important part of the industry of both Britain and the Netherlands in the centuries that followed. In Britain they were mostly used for milling grain; in the Netherlands many were used for dewatering polderland.

By the 18th century windmills were one of the highest forms of technology. They could produce 30-40 kW of power (which is about the same as the power of a small motor car). With the advent of steam power and later the internal combustion engine in Europe, the incentive to develop windmills disappeared. Instead, windmill development continued in the USA. In the mid 19th century settlers were moving into the Great Plains where there was a shortage of fuel and transport was difficult. With the need for water and the steady, regular wind across the Great Plains, windmills were an ideal technology. By the 1880's the familiar all-steel American multi-bladed farm windpump had evolved. It looked not much different from many that are still in production today.

Past experience which has led to the adoption of present-day windpump designs

Most modern efficient windpumps are horizontal axis, multi-bladed (see Figure 1). Other designs have been tried in the past and have proved less satisfactory for water pumping. They are briefly described below:

Figure 1:

Multi-bladed horizontal-axis windmill (side view)

2- or 3- bladed horizontal-axis windmills are used for electricity generation. They are not suitable for water pumping directly because

1. they cannot produce enough torque to start a piston pump working; and

2. they rotate too quickly to directly drive a reciprocating pump. These wind turbines are also more difficult to manufacture owing to the precision engineering needed.

However they could be used indirectly for water pumping by generating electricity and using this to drive electric pumps. This option is expensive but may be suitable for some locations or when a large amount of power is needed.

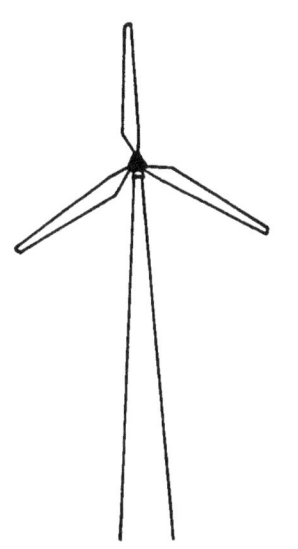

Figure 2: Three-bladed horizontal-axis windmill (side view)

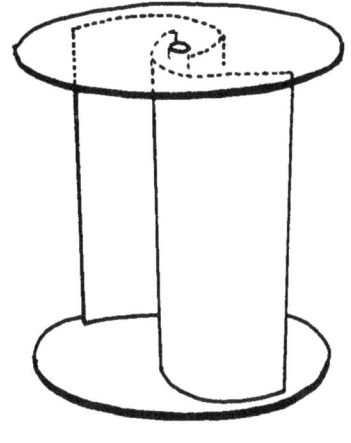

Figure 3: Schematic diagram of Savonius rotor (side view)

Savonius rotors are turned by the drag force of the wind mostly, rather than the lift force. They are therefore inefficient and turn very slowly. (See Sections 1.3 and 2.1 for explanations of drag and lift forces).

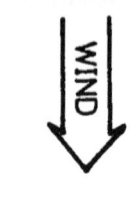

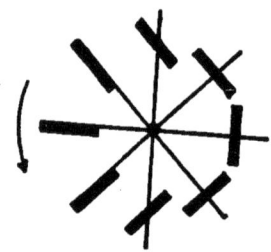

Panamones are turned entirely by the drag force of the wind. They suffer the same disadvantages as Savonius rotors.

Figure 4: Panamones (plan views)

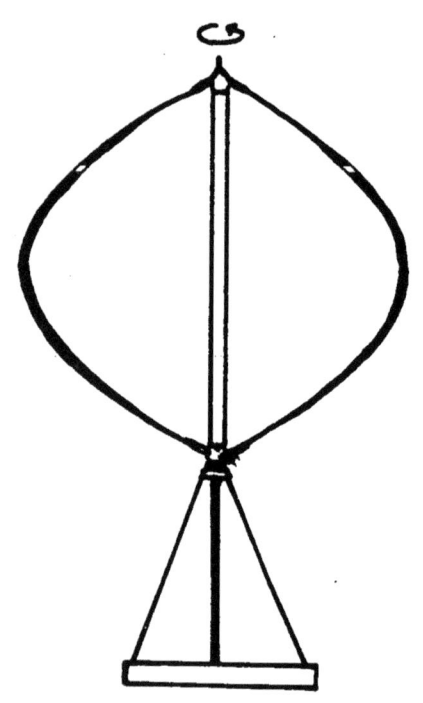

Cross flow or Darrieus wind turbines are attracting some attention at present. However they are unsuitable for water pumping because they cannot normally self-start. Even if they are modified to enable them to self-start they cannot produce sufficient torque to start a pump. They are difficult to protect from storm damage and have not yet been manufactured more cheaply than horizontal-axis rotors.

Figure 5: Schematic diagram of Darrieus wind turbine (side view)

The remainder of this Handbook concentrates on multi-bladed horizontal-axis windpumps as the only practical, commercially-available technology for water pumping at this time.

1.3 The wind energy resource

Many areas of the world are sufficiently windy for windpumps to be a realistic option for pumping water. Figure 6 shows a contour map of the average annual wind speeds for the world (Reference 1). It must be remembered that, in general, the basic requirement for wind to be a reasonable option for water pumping is that the average wind speed in the most critical month (i.e. the month where the demand for water is greatest in relation to the wind energy available) is greater than 2.5 m/s (6 mph or 5 knots). The wind will vary from day to day and month to month. It is important that there is sufficient wind available <u>throughout</u> the period when water is needed. If the water is for irrigation it may be needed for only a few months, but if the water is for domestic consumption, there must be sufficient wind all year. It is advantageous to have reliable windspeed data for at least a year to decide firstly whether a windpump is a possible option, and secondly what size of windpump to use, and how much water storage is needed.

This section briefly explains how to determine the energy available from the wind if the wind speed is known. Section 3.1 will explain how, where and how often to measure wind speeds.

The effect of wind speed

The power in the wind, and therefore its energy, is proportional to the cube of the wind velocity. This means that as the wind speed increases, the power available increases much faster. For example, in very light winds there is about 10 W/m^2 whilst in hurricane-level winds there is about 40,000 W/m^2. This extreme variability of the wind power strongly influences most aspects of system design, construction, siting, use and economy. In comparison, the solar energy resource is much less variable, there being about 100 W/m^2 in weak sunshine and 1000 W/m^2 in the strongest sunshine.

The equation describing the power in the wind is:

$$\boxed{\text{Power in W}} = \boxed{1/2} \times \boxed{\text{Density of air in kg/m}^3} \times \boxed{\text{Cross-sectional area in m}^2} \times \boxed{(\text{Velocity})^3 \text{ in m/s}}$$

The effect of air density

The density of the air affects the energy available to a very much lesser extent than the wind velocity. However it should not be ignored. The density of the air is affected by:

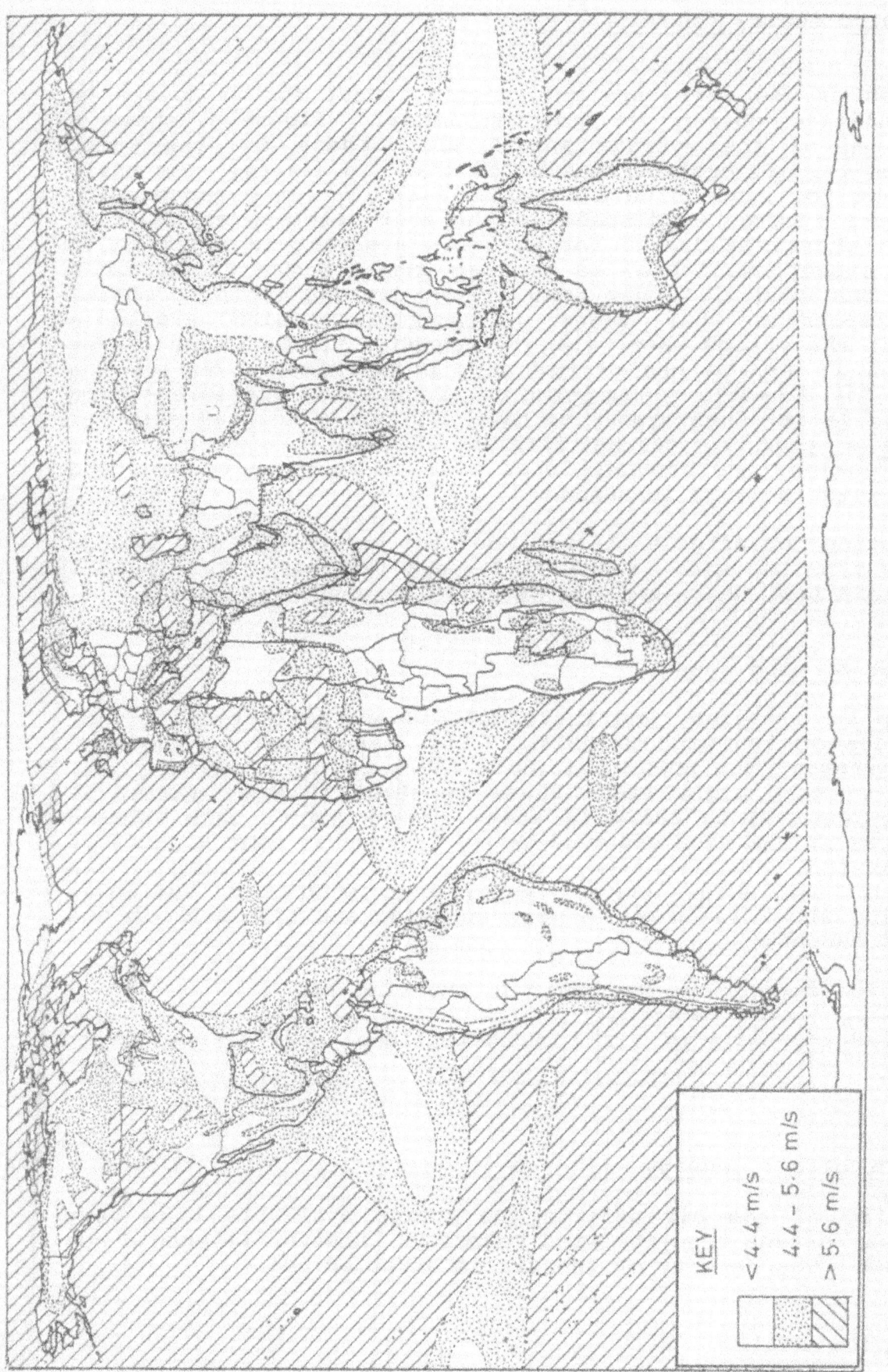

Figure 6: Global annual average wind speeds. (Redrawn from World Meteorological Society data in WMO Technical Note on Wind Energy. Reference 1)

Note - Very large local variations occur in wind speeds. This map should not be used for windpump siting. It is included to give a general indication only.

KEY
< 4.4 m/s
4.4 - 5.6 m/s
> 5.6 m/s

1. altitude
2. temperature
3. atmospheric pressure.

The effects of temperature and atmospheric pressure are very small compared with altitude and may therefore be ignored. Allowance should be made for altitude, however. For example at an altitude of 1000 metres the energy available from the wind at a given wind speed is reduced by 11%.

Table 1 gives the altitude correction factors which should be applied to the air density in order to calculate the available wind energy. Air density at sea level is 1.2kg/m^3. Figure 7 shows the same information graphically on an energy-versus-wind speed graph.

Altitude in metres above sea level	0	1000	2000	3000
Air density correction factor	1.00	0.89	0.78	0.69

Table 1: Altitude correction factors for air density

Example: To find the air density at 2000 m

$$\boxed{\text{Air density at 2000 m}} = \boxed{\text{Air density at sea level}} \times \boxed{\text{Correction factor}}$$

Air density at 2000 m = 1.2 x 0.78

= 0.94 kg/m^3

However, the wind tends to blow at higher speeds at higher altitudes. This often more than compensates for the loss due to reduced air density.

Energy availability

Only part of the energy in the wind is available for use. To extract all the energy would require bringing the wind to rest which is impossible. The available energy is extracted by slowing down the wind and using some of its kinetic energy. The maximum amount of energy that can, even in theory, be physically extracted from the wind is 59.3% of the total available. In practice wind rotors are not perfectly efficient. Good ones will be able to extract 25-40% of the total kinetic energy.

To find the amount of energy available from the wind

The graph in figure 7 may be used to find out the amount of energy that is obtainable from the wind by a well-designed windpump in a typical wind regime.

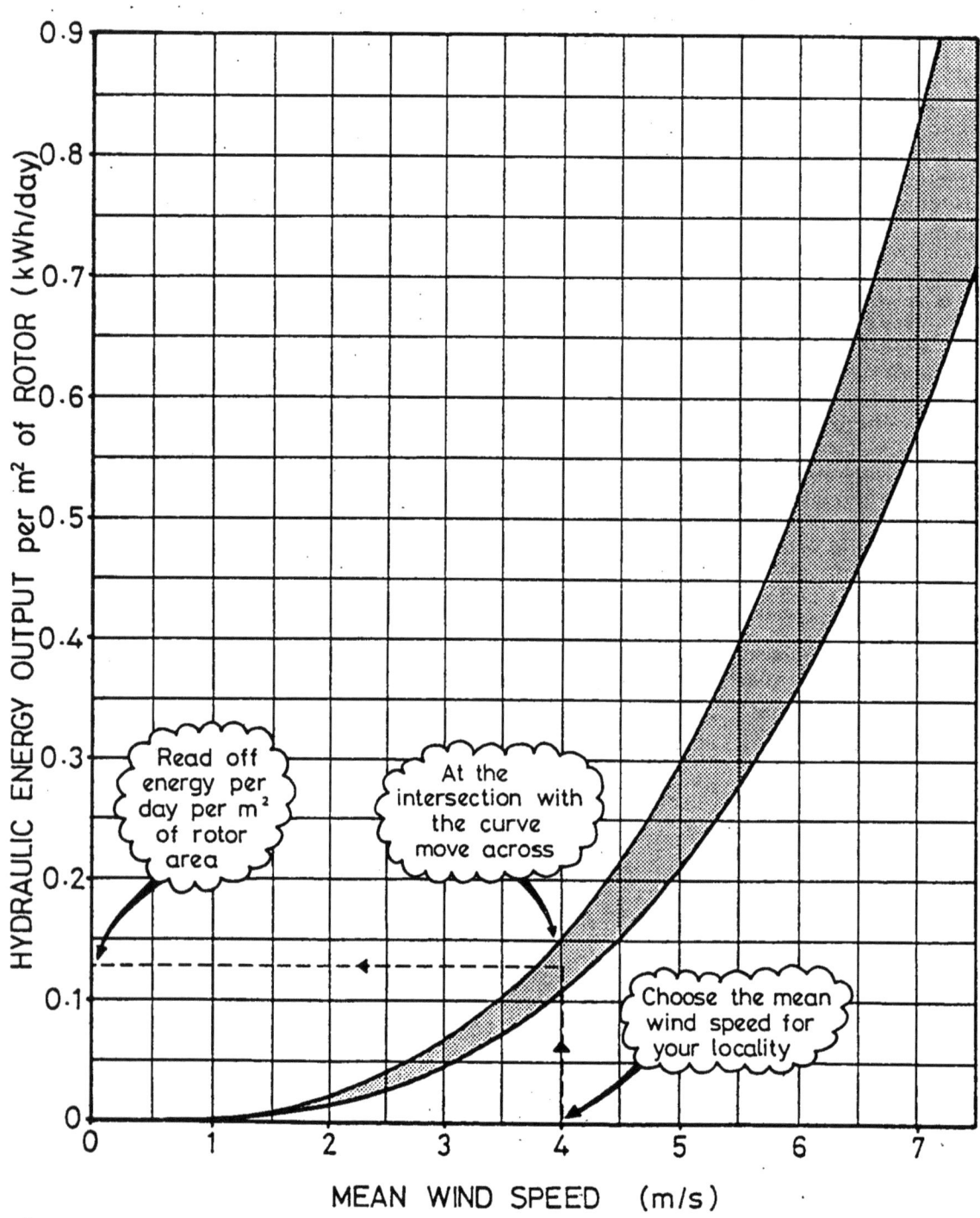

Figure 7: Approximate hydraulic energy output of a windpump rotor for various wind speeds

On the horizontal wind speed axis, choose the average wind speed for the locality in which you are interested. (Note: methods of measuring the average wind speed are described in Section 3.1). Move vertically up the graph until you reach the curve. The width of the curve is due to variation with altitude. This variation is small compared with uncertainty in windspeed. Now move across to the vertical axis to obtain the power extractable per square metre of rotor area.

An example is shown on Figure 7 for a wind speed of 4 m/s.

1.4 Choices of energy resource

There are numerous ways in which the energy for water pumping may be supplied. These are:

    Diesel or gasoline engine
    Electric grid extension
    Hydroelectric generator or hydroturbine pump
    Solar energy (using photovoltaic cells)
    Windmill
    Biogas combustion engine
    Wood- or charcoal-burning engine
    Human muscles
    Animal muscles.

In choosing the most suitable method from this list for a particular application several factors must be taken into consideration:

1. The availability of the resource -
    Is there enough energy when it is needed? For example, is the solar irradiation sufficient throughout the year to power a domestic water pump?
    Can animals provide the amount of energy required to lift the water?

2. The pumping head and the type of water source -
    The types of pumping system suitable for an open water source and a low head are different from those suitable for boreholes and medium or high heads.

3. What are the maintenance requirements? -
    Are there local people who can carry out maintenance?
    Are spare parts available locally?

4. What is the likely reliability of the energy resource conversion?
    Is this acceptable?

5. For the options remaining, which is the most economic? Section 4.4 explains how to use life-cycle costing, and in particular how to cost a windpump.

The reader is referred to Section 4.2 and References 2,3 and 4 for further information about pumping systems other than wind.

When considering whether a windpump is the best option there are a number of possible advantages that should be borne in mind:

1. Windpumps should require very little maintenance, and be very reliable. These two aspects make them attractive for remote places where skilled supervision may not be available.

2. Windpumps have a long life expectancy. (Possibly more than 20 years).

3. Windpumps are mechanically simple and suitable for small-scale manufacture.

4. Windpumps are less liable to theft than small engines or photovoltaic arrays.

Table 2 shows which countries are known to be manufacturing windpumps in 1986. The total present world population of working windpumps is estimated to be between 0.5 and 1.0 million plus about 0.3 to 0.4 million specialized paddy irrigation windpumps said to be in use in China.

Argentina	New Zealand
Australia	Pakistan
Brazil	Peru
Canada	Philippines
Cape Verde Islands	Portugal
China	Senegal
Colombia	South Africa
Denmark	Sri Lanka
Finland	Taiwan
France	Thailand
Germany (West)	UK
India	USA
Italy	Zimbabwe
Kenya	
Netherlands	

Table 2: Countries known to be manufacturing windpumps in 1986

CHAPTER 2: WINDPUMP DESIGN - STATE OF THE ART

2.1 Principles of wind energy conversion: Lift and drag

When the wind is blowing it exerts two types of force, lift and drag, on the objects in its path. The difference between lift and drag is illustrated by sailing boat sails.

DRAG A spinnaker sail uses the drag force of the wind. The sail fills with wind like a parachute and pulls the boat along.

Drag force acts in the same direction as the wind.

Figure 8: A boat propelled by the drag force of the wind

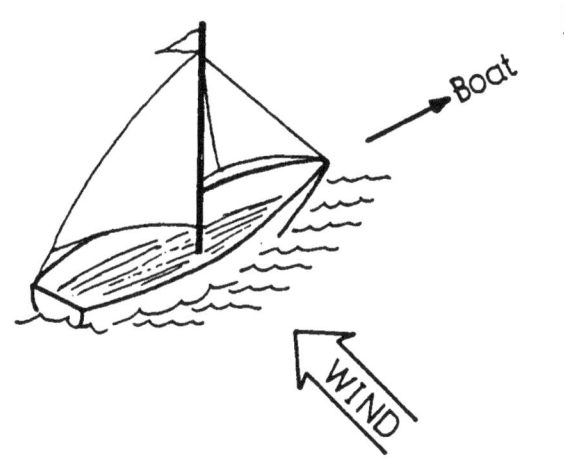

LIFT Bermuda rigging uses the lift force of the wind. The triangular sail deflects the wind and allows the boat to travel across the wind.

Lift force is perpendicular to the direction of the wind.

Figure 9: A boat propelled by the lift force of the wind

The relative sizes of the drag and lift forces depend entirely on the shape of the object. Streamlined objects experience much smaller drag forces than blunt objects.

Generation of lift always causes a certain amount of drag force.

Good aerofoils can have lift 30 times greater than drag. Figure 10 illustrates this. Lift devices are inherently more efficient than drag devices.

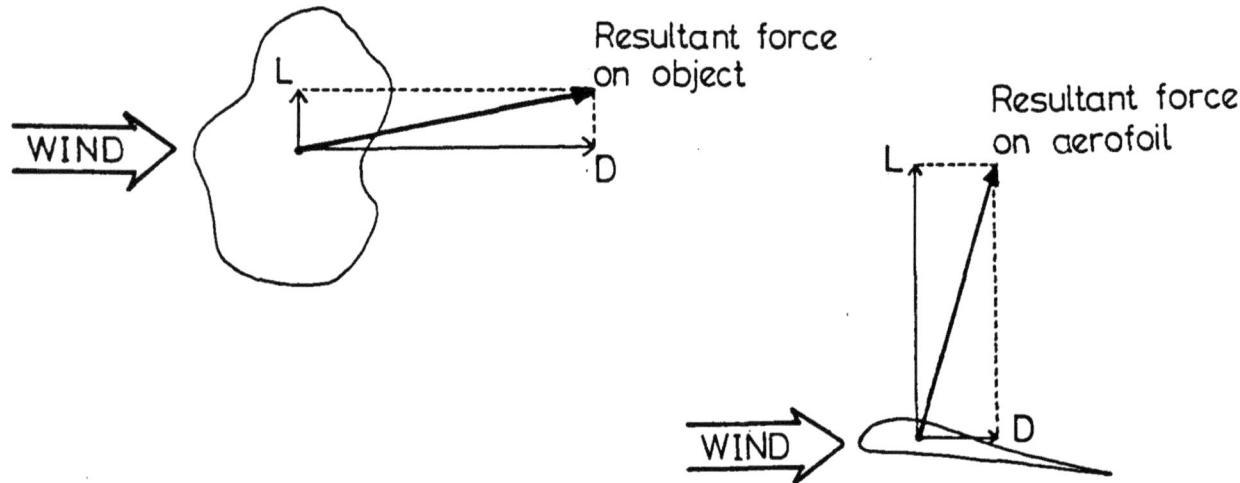

Figure 10: The relative sizes of lift and drag forces for blunt and streamlined objects

Lift forces are produced by causing a difference in the velocity of the airstream flowing either side of the lifting surface. On the surface where the air is flowing faster, the pressure drops. This creates a pressure difference across the surface. The pressure difference produces a force acting from the high pressure side towards the low pressure side.

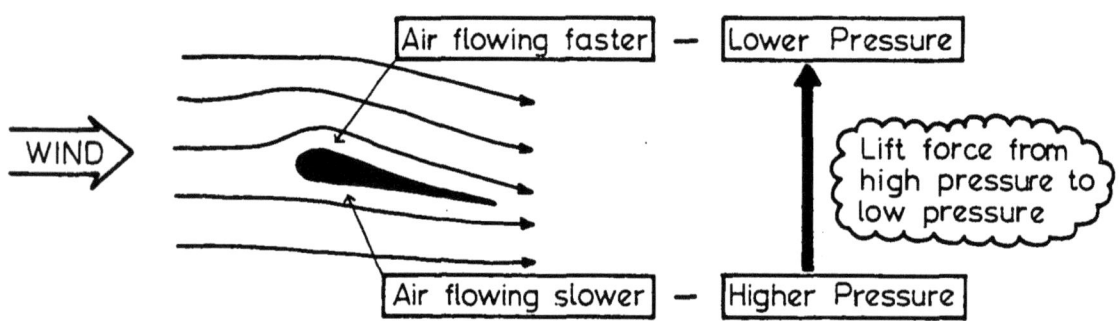

Figure 11: Generation of lift by an aerofoil

The methods by which lift and drag forces are created and do useful work are outlined in the section that follows.

2.2 Rotor design

Horizontal-axis multi-bladed rotors are used in the vast majority of successful windpumps, as stated in Section 1.2. This section is restricted to this type of rotor, although occasional reference will be made to the other types illustrated on pages 3 and 4.

In order to understand the effects of differences in rotor design it is useful to describe how the blades of a rotor react to the wind, and to define some of the standard design parameters.

Pitch

The blades of a rotor are curved so that they deflect the wind, as illustrated in Figure 11. The lift force created causes the rotor to rotate. In order to generate the maximum amount of lift, the blades must be set at an appropriate angle to the wind, called the pitch (see Figure 12). Since the tips of the blades travel faster than points nearer the axis, the angle of the wind "seen" by the blade changes with the radius. This is illustrated in Figure 13.

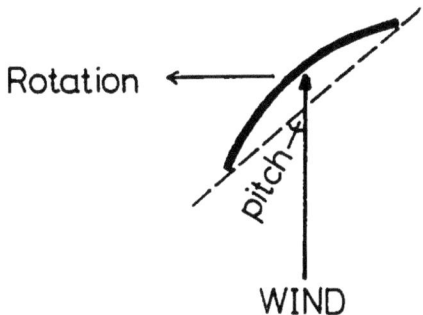

Figure 12: Schematic diagram showing angle of pitch of a rotor blade

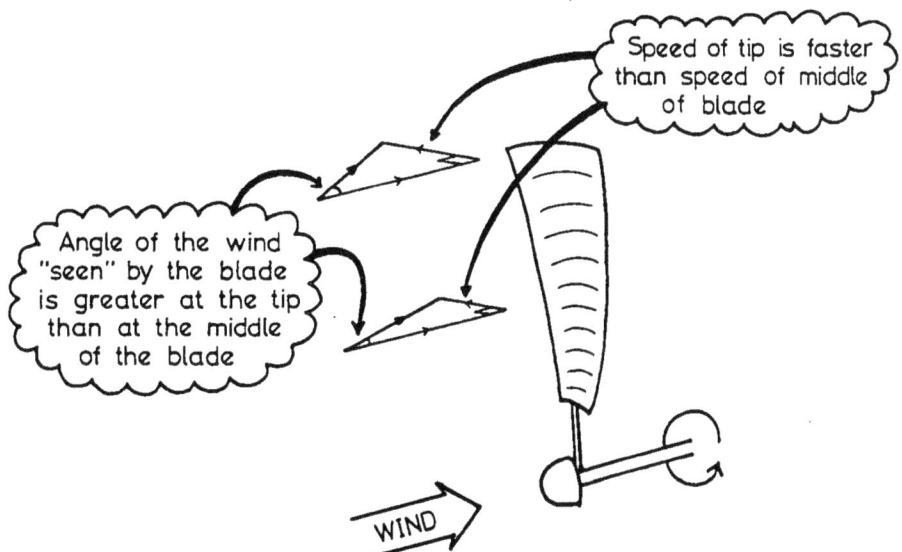

Figure 13: Change of blade pitch with radius

The rotor is most efficient if this angle "seen" by the blade is as large as possible without being so large that the rotor stalls. To make the angle large all the way along the blade it must be twisted. For the same reason, a rotor designed to rotate fast, such as a 2- or 3- bladed wind turbine, has its blades set at a smaller pitch.

Solidity

This is usually defined as the percentage of the circumference of the rotor which is filled by rotor blades. For example, if a 6 m diameter rotor has 24 blades, each 0.35 m wide, its solidity is calculated as:

$$\text{Solidity} = \frac{24 \times 0.35}{\pi \times 6} \times 100$$

$$= 45\%$$

It is, in effect, the fraction of the swept area of the rotor which is filled with metal.

The general equation is:

$$\boxed{\text{Solidity \%}} = \boxed{31.8} \times \boxed{\text{Number of blades}} \times \boxed{\text{Blade width}} \div \boxed{\text{Rotor diameter}}$$

The greater the solidity of a rotor the slower it needs to turn to intercept the wind. A 2- or 3- bladed wind turbine (see Figure 2) has a very low solidity and therefore needs to rotate quickly to intercept the wind. Otherwise a lot of wind energy would be lost through the large gaps between the blades. The typical solidity of a multi-bladed rotor for water pumping is 40 to 60%.

Tip-speed ratio

This is the ratio of the speed of the blade tips to the speed of the wind. For example, if a 6 m diameter rotor is rotating at 20 rpm (revolutions per minute), and the wind speed is 4 m/s, the tip-speed ratio of the rotor is given by:

$$\text{Tip speed ratio} = \frac{(\pi \times 6 \times 20)/60}{4}$$

$$= 1.6$$

The general equation is:

$$\text{Tip-speed ratio} = 0.052 \times \text{Rotor diameter in m} \times \text{Rotation speed in rpm} \div \text{Wind speed in m/s}$$

If the rotor is rotating faster than the wind speed it will have a tip-speed ratio of greater than one. Conversely if it is rotating more slowly than the wind speed it will have a tip-speed ratio of less than one. Rotors which rely on drag forces to turn them, such as the panamones briefly described in Section 1.2, can never rotate faster than the wind speed and will always have tip-speed ratios of less than one. Two- and three- bladed wind turbines which rotate very fast have high tip-speed ratios of 3 to 10. Multi-bladed rotors suitable for wind pumping generally have tip-speed ratios between 1 and 2. Every rotor has an optimum tip-speed ratio at which it is operating at its maximum efficiency.

Performance coefficient

The performance coefficient of a rotor is the fraction of wind energy passing through the rotor disc which is converted into shaft power. This is a measure of the efficiency of the rotor and it varies with the tip-speed ratio. Typical performance coefficient versus tip-speed ratio curves for rotors of varying solidity are shown in Figure 14. Each type of rotor has a unique performance coefficient versus tip-speed ratio curve.

Torque

Torque is the turning force produced by the rotor. It depends on the solidity and tip-speed ratio of the rotor. High-solidity rotors with low tip-speed ratios (such as the multi-bladed windpump rotors) produce much more torque than low-solidity, high-speed machines (such as wind turbines). Figure 14 illustrates this. The important features to note are that the higher speed machine has a slightly higher maximum performance coefficient but a low starting torque. Conversely the high solidity rotor produces a high starting torque but has a slightly lower maximum performance coefficient.

The choice of rotor depends on the load characteristics. A positive displacement pump, such as the piston pumps used in boreholes, demands a higher starting torque than running torque and therefore a high-solidity rotor is almost essential unless some method of unloading the rotor to help it start is included. However, electricity generators need little torque to start them turning and they need to be driven at high speed, so generally a high-speed, low-solidity rotor is used for this type of load.

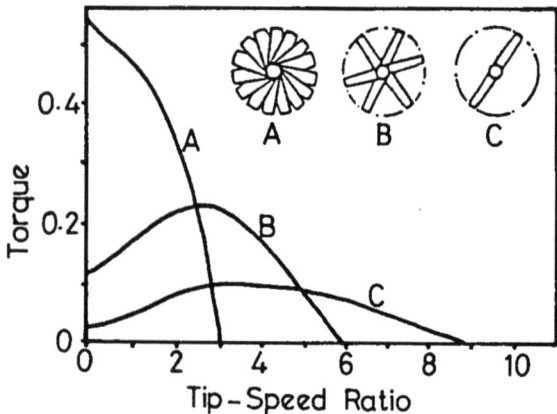 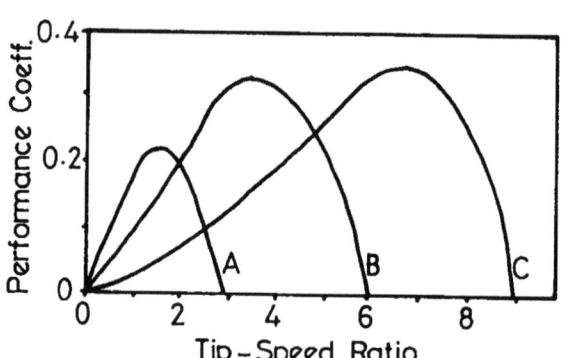

Figure 14: Typical torque versus tip-speed ratio and performance coefficient versus tip-speed ratio curves for rotors of varying solidity

Positive displacement pumps, which are invariably used with windpumps, need a fairly high torque to start, but will then continue to run with a lower torque. The rotor of a windpump will always operate at a speed such that the torque produced exactly matches the torque required by the pump. For this reason the torque characteristics of a windpump are important. In order to produce a high starting torque a high-solidity rotor is needed. This is why windpumps are almost always designed with high-solidity multi-bladed rotors.

For a reciprocating positive displacement pump approximately three times as much torque is needed to start it than to keep it running. This means that even if a windpump will operate at low wind speeds, it will need a gust of higher wind speed to actually start it. For example, a windpump may need 5m/s wind speed to start it but will continue to run if the wind speed drops to 3 m/s. If the wind speed drops lower than 3m/s the pump stops. If the wind speed then increases to 4m/s the windpump would not operate. It would only start again if the wind reached 5m/s. This is a very important point for which allowance must be made in windpump sizing. It is illustrated in Figure 15.

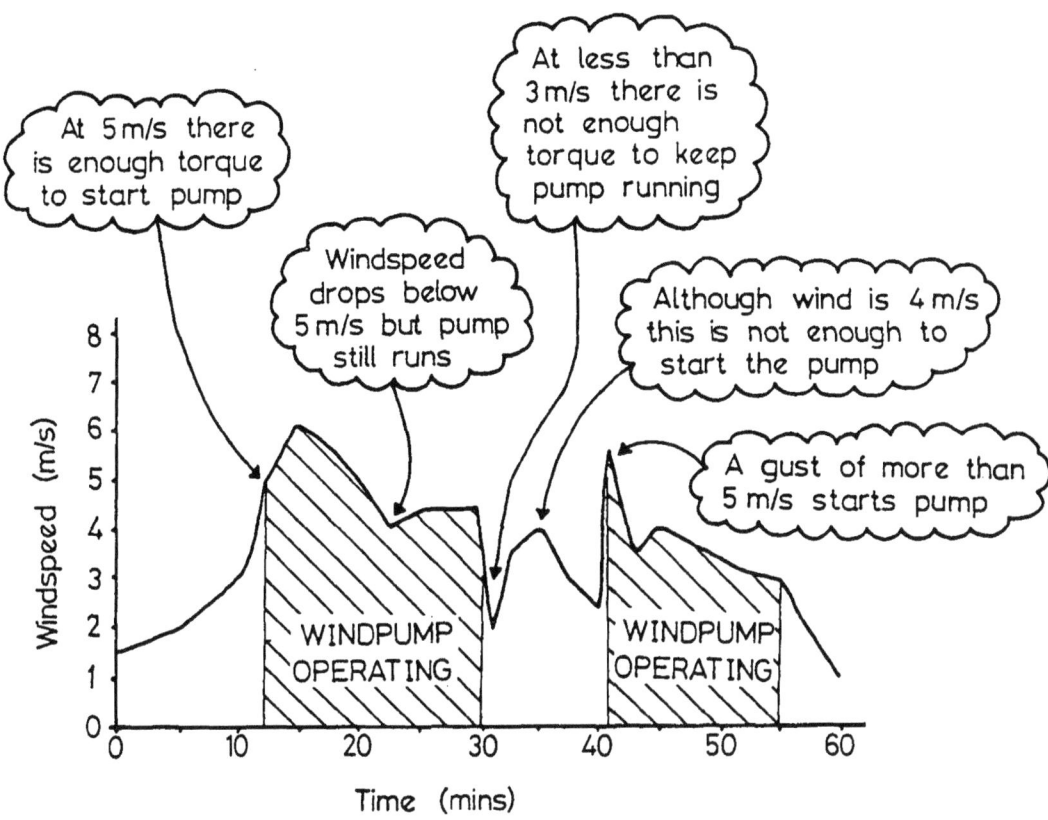

Figure 15: Schematic diagram to illustrate the effect on windpump operation time of the high starting torque

2.3 Pump types

Pumping from boreholes and deep wells requires different pump types from surface water pumping. Borehole and deep well pumping use submersed pumps. These pumps are generally from one of two major categories of pumps; centrifugal, or positive displacement. Surface water pumping, on the other hand, uses surface-mounted pumps with suction lift. These pumps are of many different types, but the most common power-driven ones are centrifugal, axial flow or mixed flow suction pumps.

Most of the remainder of this section relates to positive displacement pumps for boreholes and deep wells, since these are most commonly used with windpumps. Some attempts to use centrifugal pumps have been made but they have mostly not succeeded. Further information about surface suction pumps can be found in Reference 2.

Centrifugal pumps always have a rotary drive whereas positive displacement pumps may have either a reciprocating or a rotary drive. Centrifugal pumps are rarely used for windpumps for two reasons. Firstly they require relatively high rotational speeds of the order of 1000 rpm or more. Such high speeds require a multi-stage transmission which, owing to its complexity, makes manufacture and maintenance more difficult. Secondly they will not operate satisfactorily over a wide range of rotational speeds. At speeds less than about 1000 rpm they will not produce sufficient lift to raise any water and at speeds greater than about 3000 rpm their efficiency has decreased so much that once again they will not produce sufficient lift to raise the water.

The only situation in which centrifugal pumps are likely to be successful is in wind-electric applications where the wind energy is converted to electricity which is used to drive pumps.

Positive displacement pumps may have either a reciprocating or a rotary drive. (Reciprocating positive displacement pumps are often known as piston pumps). Both these types are used for windpumps, but they require different transmission arrangements (see 2.4). Figures 16 and 17 illustrate the principles of operation of the two types of pump.

In the piston pump, the piston is driven up and down by the pump rods. As the piston moves down the piston valve opens, allowing water to pass from below the piston to above it. As the piston moves up, the piston valve shuts and the foot valve opens. The water above the piston is lifted up and pushed through the delivery, whilst at the same time the suction created below the piston draws water in through the foot valve. This cycle is repeated with water being delivered on the up-stroke.

In the rotary positive displacement pump the rotor acts like a screw which rotates against the rubber stator. The particular shapes of the rotor and stator are such that they form a helix of cavities which are filled with water at the bottom of the cylinder and then driven upwards by the rotary movement. This type of pump is sometimes called a progressive cavity pump.

Piston pumps down wells or boreholes operate at relatively slow speeds of 1-50 strokes per minute. They should not be operated faster than this because there is a danger of seriously damaging the pump rods from compression forces. These stroke speeds are similar to the rotational speeds of large windpump rotors, so that windpump rotors of more than about 4 to 5 m diameter can be directly coupled to piston pumps without any need for gearing to reduce or increase the speed. Small windpump rotors rotate faster than large ones and generally need gearing down. The pumps will operate efficiently over their complete range of speeds, which makes them very suitable for windpumps. Their only major disadvantage is that they require a much greater torque to start them than they do to keep running.

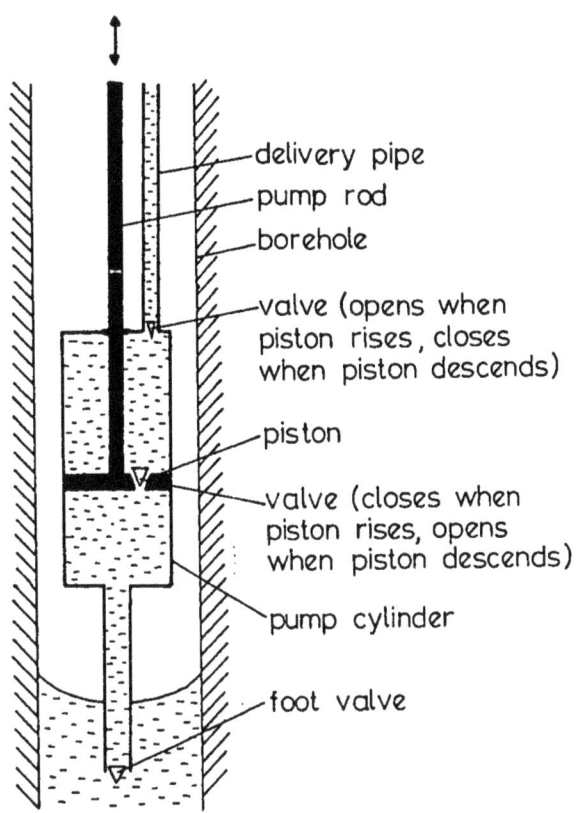

Figure 16: Schematic diagram of a reciprocating positive displacement pump (piston pump)

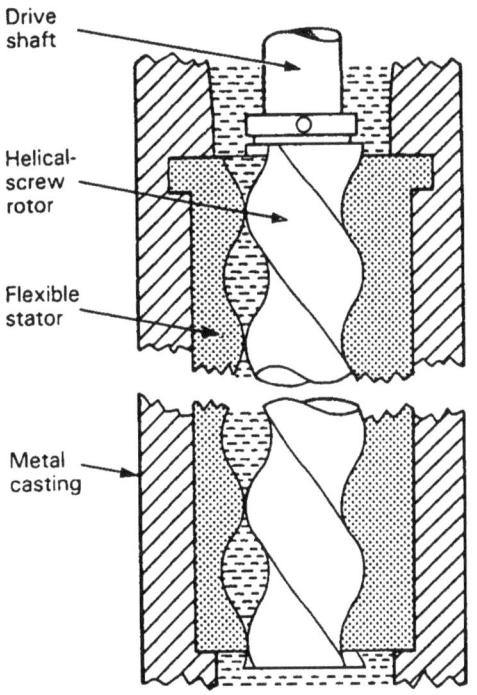

Figure 17: Schematic diagram of a rotary positive displacement pump (also known as a progressive cavity or 'Mono' pump)

Rotary positive displacement pumps share many of the same advantages as piston pumps for windpumping. They have the additional advantage that they are not subject to such large impulse forces as piston pumps and may therefore be made from cheaper, lighter components. They suffer the same problem of starting torque being larger than running torque, although for different reasons. When a rotary positive displacement pump is not in use, the rotor tends to stick to the stator, and a large force is needed to overcome this sticking when starting the pump from rest. As the pump becomes older and the stator wears, the sticking force lessens, but the pump efficiency drops because water can flow back down between the stator and rotor. Recent developments in progressive cavity pump design by one manufacturer have greatly reduced these problems.

Both reciprocating and rotary positive displacement pumps have similar head, flow and efficiency characteristics, as shown in Figure 18. The flow increases as the head decreases and as the speed of pumping increases. A pump will have a peak efficiency at one particular head. Therefore a pump should be chosen with peak efficiency at or near the required head. The pump efficiency is not dependent on the speed of pumping since the flow from the pump is approximately proportional to the speed. It may not be possible to size for peak effiency if the head is low, because positive displacement pumps are always of low efficiency at low heads.

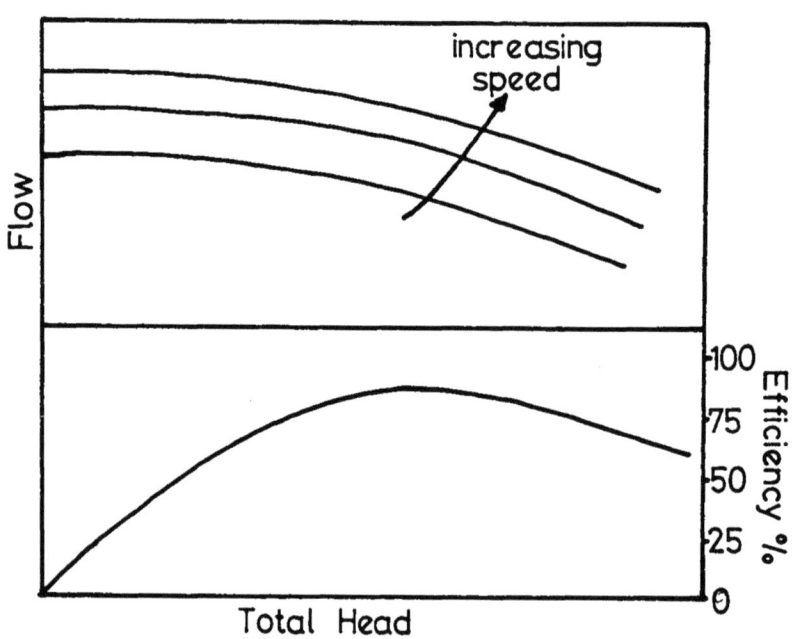

Figure 18: Typical head, flow and efficiency curves for positive displacement pumps

2.4 Transmissions, tails and towers

Transmissions

The transmission of a windpump is the part which converts the rotation of the rotor into a movement suitable to drive the pump. The precise configuration of the transmission depends on the drive required by the pump, and the manufacturer's design. For a reciprocating pump the horizontal rotation of the rotor must be converted into a vertical up-and-down movement whilst for a rotary pump the horizontal rotation of the wind rotor must be converted into a vertical rotation.

For some types of pumps the speed of rotation of the rotor is either too fast or two slow and speed gearing is needed in the transmission. For example small rotors of less than about 4 to 5 m diameter rotate too fast to drive a piston pump. Gearing not only enables the speeds to be matched but also reduces the torque needed to start the pump. However it makes the transmission more complex and more expensive. Conversely rotary pumps require gearing up as the windpump rotor does not rotate fast enough to drive the pump directly. The advantages of using a rotary pump have to be offset against the complexity and cost of the gearing.

Tails

The tail of a horizontal-axis wind rotor serves two purposes. At low and medium wind speeds when the windpump is operating it keeps the rotor facing into the wind. At high wind speeds it causes the rotor to turn sideways on to the wind (i.e. to furl) so that it is not damaged during stormy weather. The power density in high winds is very high (see Figure 7) and they can therefore do a lot of damage. Windpumps are usually designed to furl at about 10-12 m/s, since the proportion of wind at speeds greater than this is usually very small. Also the additional strength that the tower, transmission and rotor would need to continue to operate safely at high wind speeds is too costly to be worthwhile.

Figure 19 shows the most common tail mechanism for horizontal-axis windpumps. The entire rotor can swing through 90 degrees so that it is edge on to the wind. This action is entirely automatic in strong winds. The pre-load of the spring, the wind load on the rotor (i.e. its area) and the geometry of the rotor axis in relation to the tower centre line determine the wind speed at which the rotor begins to furl. Most windpumps also have a mechanism whereby the rotor can be manually furled during maintenance work.

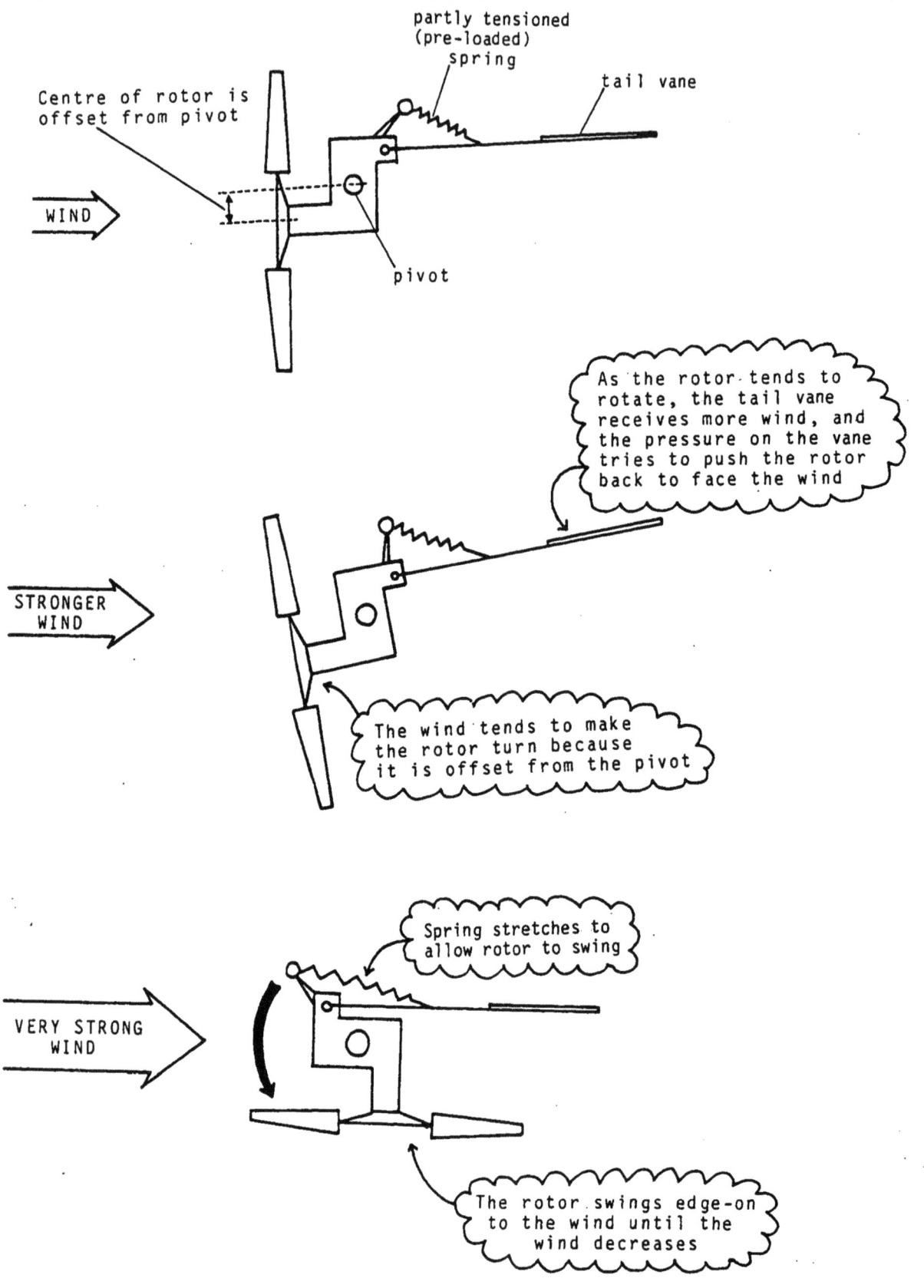

Figure 19: Schematic diagrams showing the furling action of a wind rotor in strong winds (bird's eye view)

Towers

The greater the height of the tower, the stronger it must be to endure the forces of the wind, both because of the tendency of a tall structure to bend and because the wind speed increases with height above ground. Also the greater the size and weight of the rotor, transmission and tail the stronger the tower must be.

The height of a windpump tower is usually between 10 and 20 m. Below 10 m the air is likely to be turbulent due to ground friction (see Section 3.1) whilst above 20 m the gains due to the greater wind speed are not enough to justify the extra cost. The gain in wind speed between 10 and 20 m will be 7-25% depending on the type of ground. The smoother the ground the less the gain, but the wind speed over smooth ground is likely to be higher anyway. Section 3.2 gives more information about ground roughness. A 7% gain in wind speed may seem small, but it must be remembered that the power available is proportional to the cube of the wind speed. A 7-25% gain in wind speed is equivalent to a 23-95% gain in power. The height chosen for the tower will depend on local conditions. The factors to be considered are listed on page 70.

The tower of a windpump should allow for the following:

1. The pump and rods to be readily removed from the borehole for maintenance.

2. Access to the rotor, transmission and tail for maintenance.

3. To be securely fixed to its foundations.

4. To be transported and site-assembled.

Some towers are hinged at the base to allow clear access to the borehole, and to make rotor assembly easier. For such windpumps only simple lifting tackle is required. The majority of windpumps are not hinged and therefore have to have the rotor assembled after the tower is erected, or have to have the rotor assembled on the ground and lifted into position on the tower with a crane, gin pole or derrick.

2.5 The feasibility of local manufacture

Advantages

There are many advantages of local windpump manufacture either to the country as a whole, or to individual purchasers, or to both.

1. The country does not have to use valuable foreign exchange to buy windpumps.

2. The country is more self-reliant with respect to meeting its energy needs.

3. Jobs are created in the industrial sector.

4. There are no import taxes to pay on windpumps.

5. Assistance with specification, installation and operator training is available from within the country.

6. Spare parts are relatively easily and quickly obtained.

Criteria for success

Windpumps are ideally suited to local manufacture in the majority of developing countries if the following criteria are met:

1. There is a large enough market for windpumps in the country (i.e. the wind regime is adequate and other types of water pumps are more costly).

2. The raw materials, pre-fabricated components and tools are available (e.g. paint, welding rods, bearings, etc).

3. The individual or organization starting the business has sufficient capital to invest in the necessary tooling.

Types of design suitable for local manufacture

In general designs based on fabricated steel construction are readily made in most developing countries. Castings, forgings, stampings, plastic components and special alloys should be avoided as far as possible because they demand a greater scale of manufacture to be economic than is appropriate.

Horizontal-axis multi-bladed rotors can be made by cutting, machining, bending, welding and bolting standard steel sheets and sections. A suitable transmission can also be manufactured by such techniques although some standard bearings may have to be bought ready-made. Some components of the transmission which bear substantial loads would benefit from heat treatment to improve their strength and toughness characteristics.

Suitable pumps could be either manufactured or bought ready-made. All countries have pump suppliers, and some have pump manufacturers locally.

Further information about setting up a small windpump manufacturing business can be found in Reference 5.

CHAPTER 3: SITE EVALUATION

3.1 Assessing the wind regime

In Section 1.2 the relationship between the wind speed and the power available has been explained. To size a windpump so that it provides sufficient water and runs efficiently in the local wind conditions, one or a combination of the following methods should be used:

1. Wind measurement at proposed windpump site.

2. Install a windpump and adjust it to suit the wind regime if necessary.

3. Infer the wind speeds from nearby site data.

Method 2 is often used if there are already windpumps in the area, since the experience of other users is generally a better guide that wind speed measurements.

If method 1 or 3 is used, the remainder of this section provides some guidance. Measurements should be taken as close as possible to the proposed windpump site, using one of the methods outlined in the following paragraphs.

The wind regime parameters needed

Ideally the following parameters should be measured:

1. Average annual windspeed
2. Average monthly windspeed for each month
3. Typical diurnal windspeed pattern for each month
4. Length and annual distribution of lull periods
5. Maximum gust windspeed
6. Wind direction.

In practice it may not be possible to obtain numerical data for all of these. In particular estimates of (4) and (5) may have to be made based on discussions with local inhabitants.

Ideally the measurements of the parameters should be taken at the proposed windpump site, at the proposed windpump height above ground, for at least one year. In practice this is rarely possible and data from other nearby locations is used. Allowances must then be made for differences between the measurement site and the proposed windpump site.

The differences which should be taken into consideration are:

1. The ground surface

   The rougher the ground surface the more it interferes with the wind. Rough ground creates turbulence in the layers of wind above it. The following table gives coefficients for the effect on wind speed of different ground roughnesses. Coefficients are given for two typical windpump heights above ground level. Underneath the table there is an example of how to use it.

Ground surface type	Wind speed coefficients at		
	6m	9m	12m
Smooth surface sea, lake, sand	1.40	1.45	1.50
Medium roughness ground - small bushes, etc.	0.90	0.98	1.05
Rough ground - woodland, buildings, etc.	0.50	0.60	0.66

   Table 3 : Coefficients for the effect on wind speed of different ground roughnesses

Example

   If you know that the average annual wind speed in a sandy area at a height of 6m is 5 m/s and you propose to use a windpump at a height of 9 m in a maize-growing area (medium roughness),

   $$\text{Windpseed in maize area} = 5 \text{ m/s} \times \frac{0.98}{1.40}$$

   $$= 3.5 \text{ m/s}$$

   The general equation is:

   | Windspeed in area you want | = | Windspeed in area you know | x | Coefficient for area you want | ÷ | Coefficient for area you know |

More information about selecting a windpump site and tower height is given on pages 23, 35 and 70.

2. Hills, valleys and ridges

   Windpumps are most commonly used on flat plains, and not in hilly country. This is chiefly because water is rarely available on hill tops. Also the wind regime is likely to be much more erratic. If there is sufficient water and wind in an area of low hills and a windpump is considered, the points below should be borne in mind.

   Rounded hills, crests and ridges generally experience higher but more variable wind speeds than flat ground. The wind is accelerated over the hills. However, on the downwind side of the hill there may be turbulence. The amount that the wind accelerates over a hill or ridge depends strongly on the height and shape of the hill. It is not possible to give factors for all the various hill types but Figure 20 gives some very approximate values which may be used to obtain a rough estimate.

   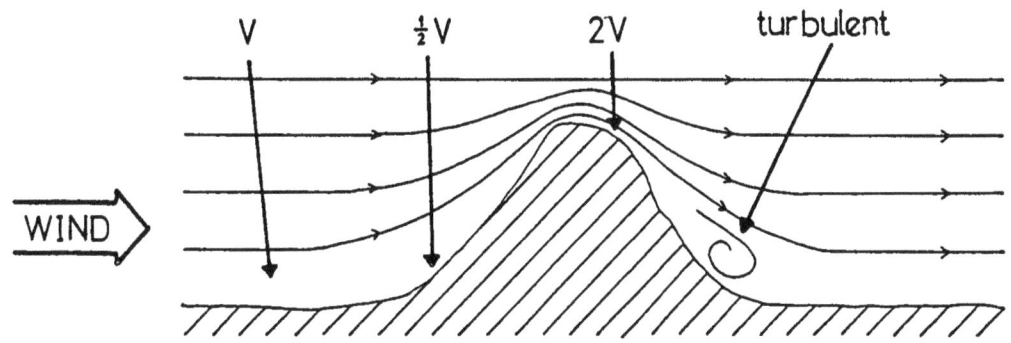

   Figure 20: Schematic diagram showing approximate wind acceleration factors over a hill

   Valleys will usually have lower and more variable wind speeds than flat ground or surrounding hills and are unlikely to be suitable windpump sites. If a valley site is being considered for a windpump Reference 6 can give some guidance.

3. Coastal areas

   Coastal areas (around large lakes as well as near the sea) are likely to have stronger winds than inland areas. The increase in wind due to a coastal location is not easily quantifiable. Wind speed data is often available at ports, harbours and lighthouses.

Sea breezes usually have a diurnal pattern due to the temperature difference between the sea and the land. They generally blow from the land to the sea during the day and from the sea to the land at night. Figure 21 illustrates this.

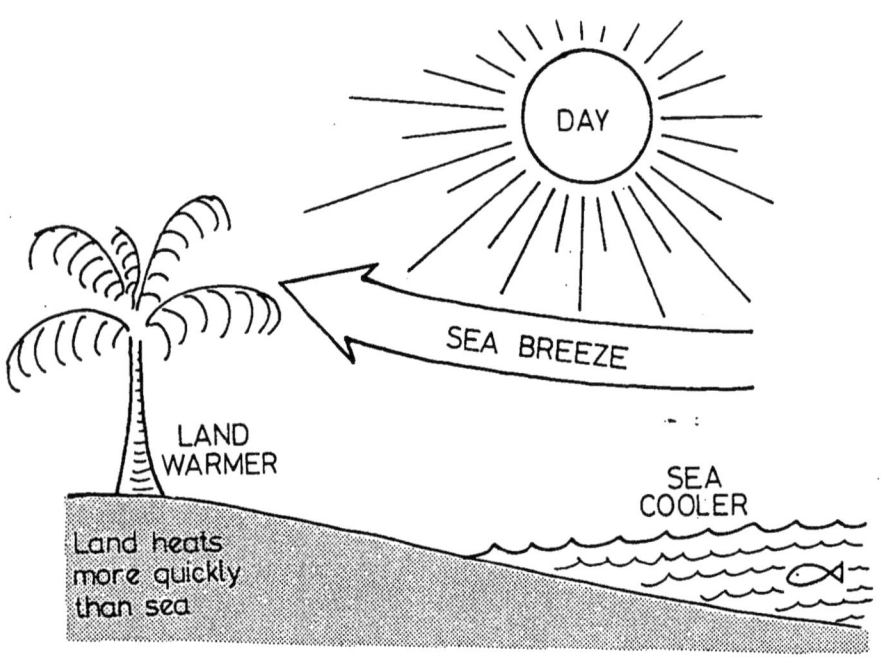

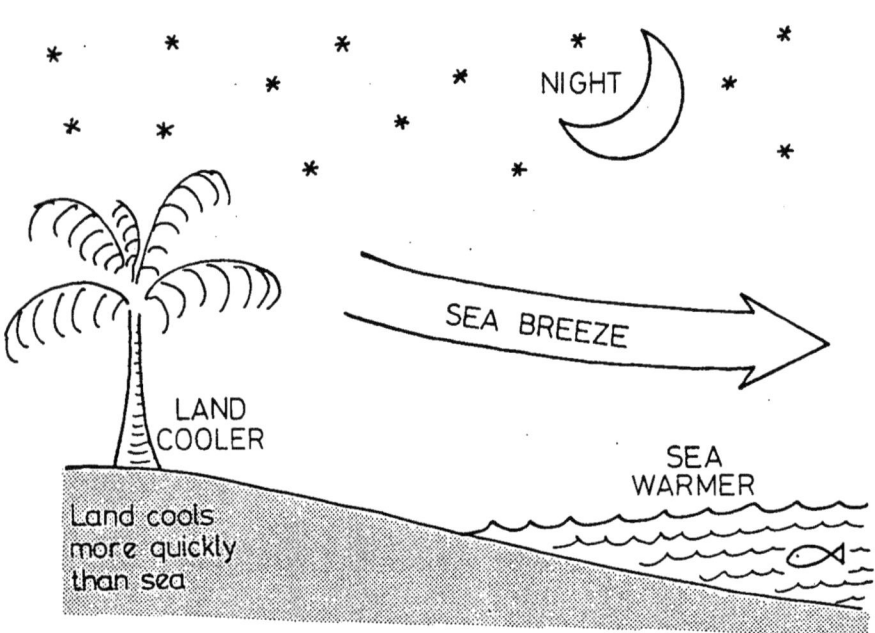

Figure 21: Sea and land breezes

4. Altitude

   As stated in (2) higher ground usually experiences stronger winds than lower ground.

Wind measurement

Various options are available for taking wind measurements depending on the time available, the budget available and the accuracy required. If there is unlimited time and money, very good accuracy can be achieved. Usually this is not the case and some sort of compromise is reached. As already stated, if windpumps are already in use in nearby areas, experience of other users is probably a better and cheaper guide to sizing the windpump than taking wind measurements. If they are not already in use, either wind measurements must be made or an estimate based on meteorological office or civil aviation data must be used. The measurements to be made are of wind speed and wind direction.

Quality of wind data

Until quite recently there was little standardisation of wind data collection except at major airports where data has had to be collected in an internationally agreed manner. Hence most airport data is collected using continuously recording anemometers located at 10 m height. This has become the world standard for meteorological stations too.

Data from rural meteorological stations on the other hand are often of doubtful accuracy. A large proportion of rural meteorological stations in many developing countries have poorly located anemometers. Such stations probably record reasonably accurate meteorological parameters such as rainfall, relative humidity, temperature or barometric pressure, because the siting requirements for those instruments are not too stringent, but the wind data is often suspect. For example, in Kenya most of the limited number of country meteorological stations use anemometers on 2 m poles and often there are buildings, trees or other obstructions close by. Unfortunately, often the entire wind energy "picture" for an enormous area may be based on the output of a single improperly sited anemometer. The data from virtually all the anemometers located on 2 m towers, are at best almost useless for wind energy assessment purposes, and at worst positively misleading. Such records should be identified, separated from "acceptable" data and rejected, even though this might result in a much smaller data base in many countries.

Although much of the wind data is unreliable, its use is unlikely to lead to the siting of windpumps in areas with inadequate winds, since wind speeds cannot normally be exaggerated by improperly sited anemometers. Therefore, most wind energy assessments are likely to be under-estimates.

Anemometers may be inaccurate, as well as poorly sited. The main cause of inaccuracy is friction in the bearings, which tends to make them rotate too slowly. Consequently if a windpump is sized using wind speed data of slightly questionable accuracy, it is likely that the windpump will be over-sized rather than under-sized. Therefore, it is important to bear in mind that there is invariably at least as much wind as appears to have been measured and there may in many cases be a lot more than was measured. The cube law relationship between power and wind speed (Section 1.3) means that a few percent under-estimate in mean wind speed can translate into quite a large under-estimate of wind energy availability.

Measurement options

1. Use wind data available from a nearby weather station and adjust it for differences in ground roughness and altitude. If the weather station data has already been summarized this method is cheap and reasonably accurate. However on many occasions there will not be a weather station sufficiently close to the proposed windpump site. As a general rule the weather station should be within 20-60 miles of the proposed site, being closer in rougher ground. In hilly countryside, a nearby weather station will be too inaccurate.

   If local weather station wind data is used, it is advisable to check the height of the anemometer and its condition.

   To be useful, the wind speed data must be summarised as monthly average wind speeds. Often it is already in this form, but if it is not it should be processed as shown in Figure 22.

2. The second option is to take limited on-site data.

   If an area is clearly windy, when considering taking on-site wind measurements it is worth comparing the cost of taking the measurements with the cost of over-sizing the windpump. In many cases it may be cheaper and possibly more effective to over-size the windpump. (If the primary use of the water is for domestic supply, the surplus from an over-sized pump could be profitably used for small-scale irrigation.)

   If the nearest weather station is far away, or is sited in terrain different from the windpump site, and it is not considered worthwhile to over-size the windpump, some on-site wind measurements must be made. These can then be compared with the weather station data. An anemometer must be installed, preferably at the intended height of the windpump rotor (see later) or alternatively at the standard height of 10 m.

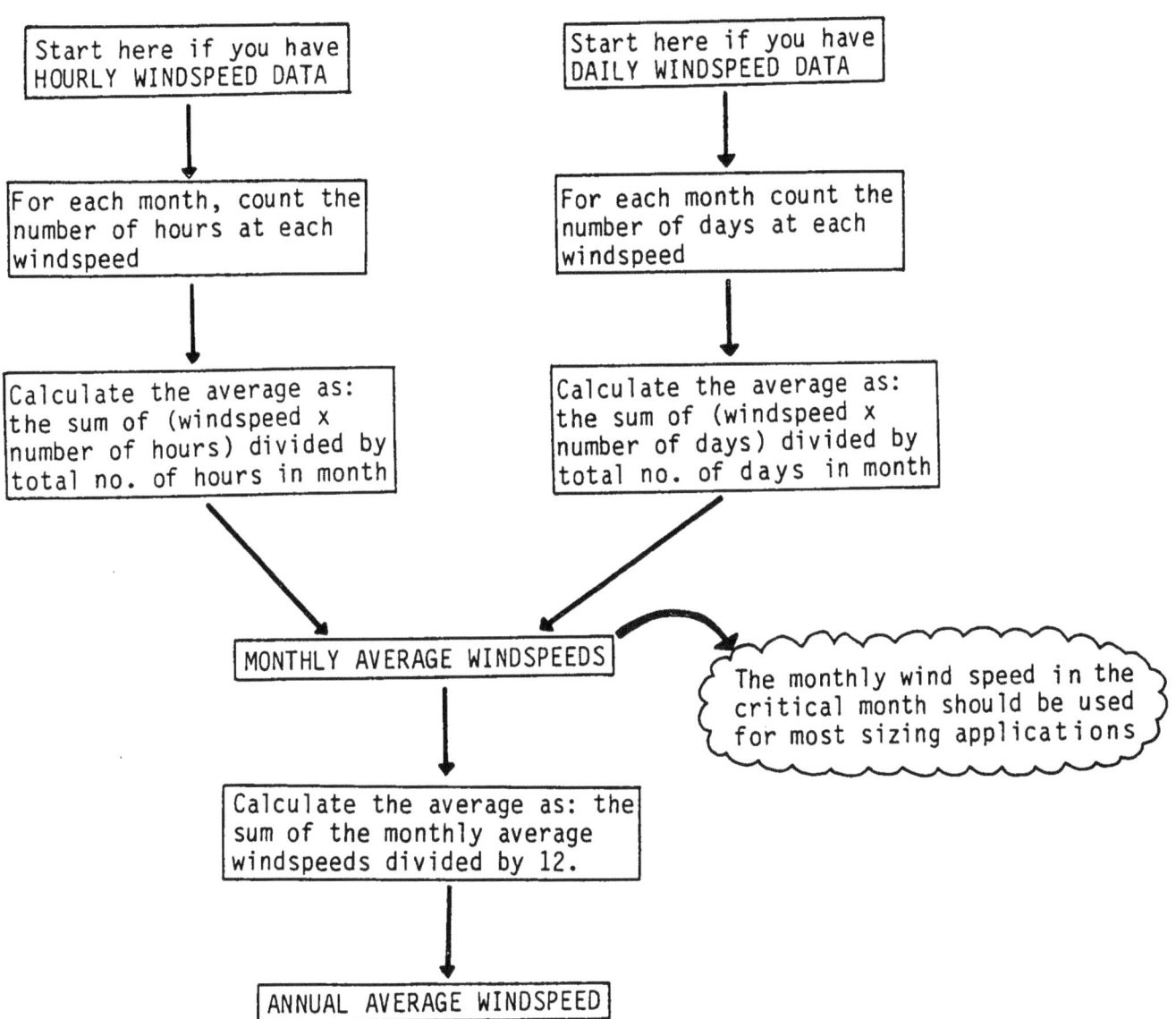

Figure 22: Flow chart outlining the steps to be taken when processing wind data

It is advisable to take on-site wind measurements hourly for at least three months. If hourly measurements are not possible, daily averages should be collected. The site data should then be compared with the weather station data for the same period, preferably using a regression analysis on the daily or hourly averages. (A local maths teacher should be able to help with regression analysis).

Knowing the difference between the site data and weather station data for three months the weather station data for other months can be adjusted to predict the wind speeds at the site.

If the site data is very different from the weather station data, the latter should not be used.

3. The third option is to take extended on-site data.

It is to be hoped that extended on-site measurements of wind speed will not be necessary, but if none of the previous methods is suitable they may have to be taken. Measurements should be taken over at least a year, in the same way as described for taking limited site data. By discussion with local inhabitants, check that no season is exceptionally windy or calm during the year of the measurements. If any season is abnormal the long-term predictions for wind may be in error. Detailed information on taking long-term wind measurements is available in Reference 7.

Choosing the windpump site

The windpump should be sited to receive the maximum amount of wind in terms of both duration and wind speed. In practice this means choosing a site which:

1. is far from obstructions (tall trees, high buildings)
2. is high up
3. receives the prevailing wind
4. avoids turbulent air streams.

At the chosen site the tower height must be sufficient that the entire rotor is in a uniform air stream (i.e. above all ground-induced turbulence). A summary of the site considerations follows.

Readers who need more detailed information are referred to Reference 6.

Buildings, tall trees and other obstructions create turbulence on the leeward side (see Figure 23). Therefore a windpump should be sited so that the rotor is well clear of the turbulent zone, either by siting it a distance of 10-20 times the height of the obstruction away from it, or by having a tower high enough to raise the rotor above the obstruction.

Although winds are generally stronger on hills, they are unlikely sites for windpumps and are therefore not discussed here. Information on this subject may be found in Reference 6.

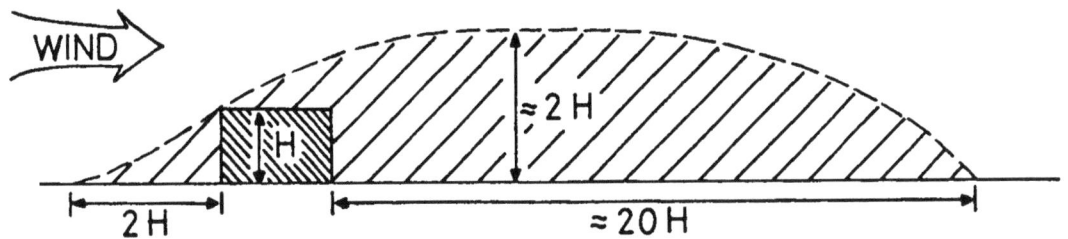

Figure 23: Area of turbulence around a building

Figure 24: Area of turbulence around trees

If there is no choice but to site a windpump in a valley or depression owing to the hyrodology of the area, or some other factor, it is important that the windpump receives the prevailing wind. In this situation it is essential to have good wind direction measurements. Since wind patterns in valleys can be quite complex, the reader is advised to seek advice from windpump suppliers, and to refer to Reference 6.

The ground, however smooth it is, reduces the wind speed close to it by friction. Figure 25 shows a typical wind speed profile up to 30 m above the ground.

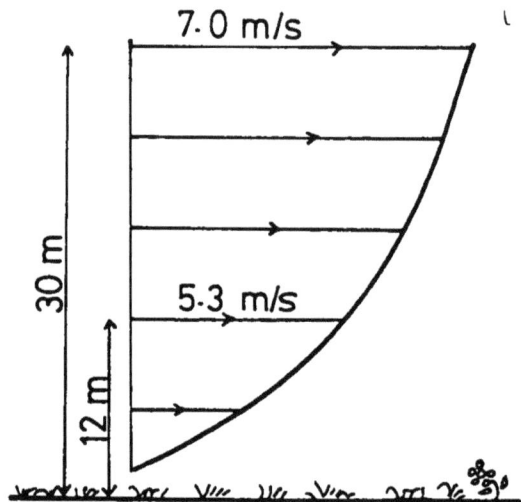

Figure 25: Effect of ground friction on wind profile

If there are crops or trees, the friction effect is greater, and there is also an area of little or no wind near the ground (see Figure 26).

Figure 26: Wind profile changes over trees, etc.

Table 3 on page 26 gives an indication of the extent to which the ground roughness affects the windspeed at a number of heights above ground level. The effect of ground friction and turbulent air streams should be minimized by mounting the windpump rotor on a sufficiently high tower. Towers are usually 10 - 20m high. It is worth noting that although the windspeed at 20 m is usually only 10 - 20% greater than the windspeed at 10 m, the power available is about 50% greater because the power is proportional to the cube of the windspeed (Section 1.3). Figure 27 shows schematically the choice of tower height to take maximum advantage of the wind.

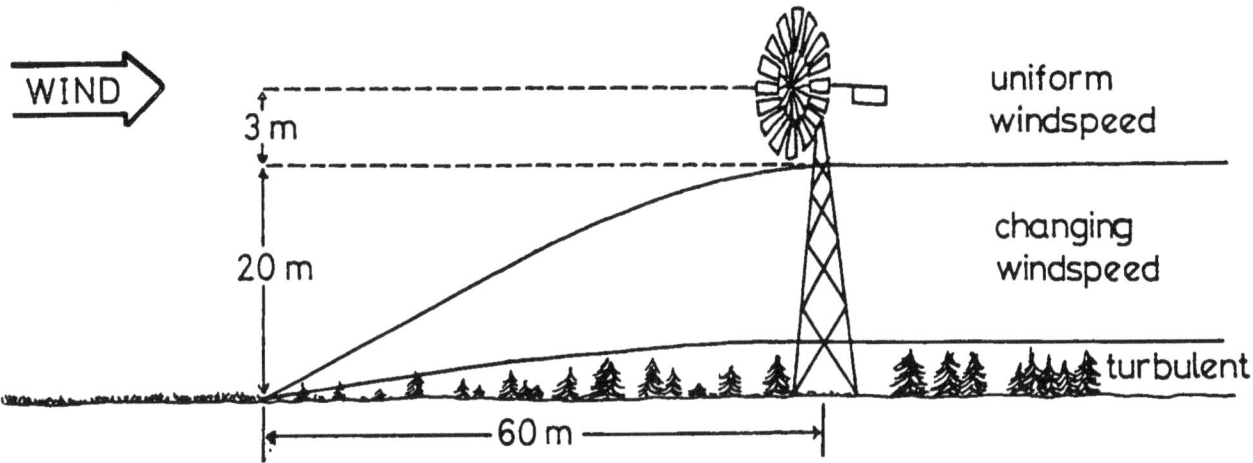

Figure 27: Schematic diagram to show selection of tower heights to achieve even wind speeds across the whole rotor

3.2 Assessing the water requirement

A windpump must be sized to meet the demand for water. This section explains how to assess the water demand for both domestic use and for irrigation. It should be noted, however, that although present day economics frequently favour windpumps for water supply, they are less commonly economic for irrigation, especially if storage is necessary. Mixed usage, where the windpump is over-sized for water supply and the excess is used for small-scale irrigation, may be worth considering.

Water for domestic use

An assessment of the domestic water demand must take account of the following:

    the present population
    the rate at which the population is increasing
    the amount of water used per person
    the number of animals
    the amount of water used per animal.

Table 4 gives typical daily water consumptions for a range of farm animals:

	Daily water need in litres
Horse	50
Cattle	40
Pig	20
Goat	5
Sheep	5
Chicken	0.1

Table 4: Daily water requirement of farm animals

The amount of water that people use increases with its availability. For example, people in the UK use about 150 litres per person per day. In developing countries, if there is a house supply, the consumption may be five or more times greater than if water has to be carried from a public water point. A WHO survey in 1970 showed that the average water consumption in developing countries ranges from 35 to 90 litres per person per day depending on the country. The long-term aim of water development is to provide all people with ready access to safe water in the quantity they want. However, in the short term the WHO recommendation for the UN Decade for Clean Water and Sanitation of 40 litres per person per day is a reasonable goal. Thus for typical village populations of 500 the water supply should provide about 20 m^3/day. There is a limit to the number of people who should be served by one centrally-located water point. In order not to cause unreasonable water collecting times and carrying distances a single water point should usually supply no more than about 250 people. It is feasible for one windpump to supply more people if the water collection points are piped to be some distance away from the windpump itself.

Since a windpump is designed to work for ten years or more, the rate of increase of the local population should be allowed for when estimating the domestic water demand. The increase in demand for water with a better supply should also be allowed for. Table 5 gives the population increase in five years and ten years for various annual growth rates.

Annual growth rate of population	% increase in population	
	in 5 years	in 10 years
1%	5	10
2%	10	22
3%	16	34
4%	22	48

Table 5: Population increase for various annual growth rates

Example: The present population of the village to be supplied by the windpump is 400 people, 200 cattle, 300 goats and 500 chickens.

Present daily water demand is

```
400 people    x 40 l  = 16000
200 cattle    x 40 l  =  8000
300 goats     x  5 l  =  1500
500 chickens  x 0.1 l =    50
                        ------
Total                 = 25550 litres
                      = 25.6 m³
```

If the population is increasing at 2% per year (and we assume that the number of animals increases at the same rate as the population).....

The daily water demand in ten years' time is

$$25.6 \text{ m}^3 + \frac{22}{100} \times 25.6 \text{ m}^3$$

$$= 25.6 \text{ m}^3 + 5.6 \text{ m}^3$$

$$= 30.9 \text{ m}^3$$

Water for irrigation

The quantity of water needed to irrigate a given land area depends on a number of factors, the most important being:

1. the crops
2. the climate
3. the type and condition of the soil
4. the topography of the land
5. the field application efficiency (i.e. the proportion of the water applied which is actually used by the plants)
6. the efficiency of the water conveyance method (i.e. some water will be lost through leaks in hoses, etc).

The quantity of water required will not be constant from month to month. At some times of year, the rainfall may be sufficient to provide all the water needed by the crops. If different crops are grown in different seasons, their water requirements will be different. The design of an irrigation pump installation will need to take account of all the factors listed above.

The crop takes its water requirements from moisture held in the soil. Useful water for the crop varies between two levels, the "permanent wilting point" and the "field capacity" (see Figure 28). Water held by the soil between these two levels acts as a store. When the store approaches its lowest level, the crop will die unless additional water is supplied.

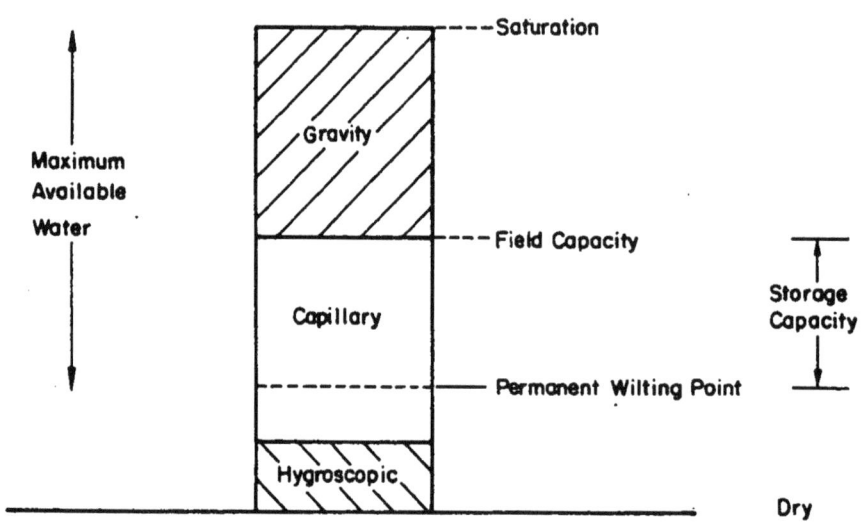

Figure 28: Soil moisture quantities

The rate of crop growth depends on the moisture content of the soil. There is an optimum growth rate when the soil water content lies at a point somewhere between the field capacity and the permanent wilting point (Figure 29). This point varies for different crops and for different stages of growth and so it is not easy to adjust the irrigation intervals so that there is optimum crop growth.

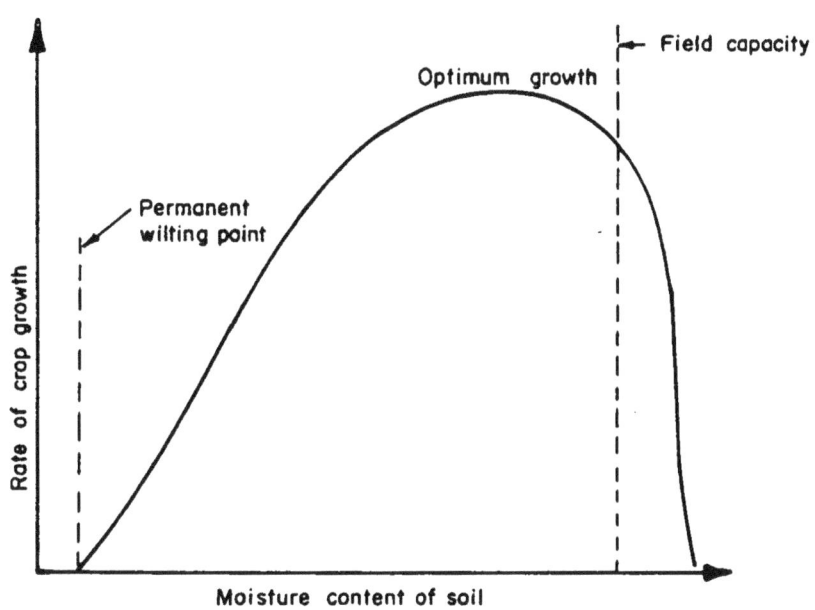

Figure 29: Rate of crop growth as a function of soil moisture content

An estimation of the quantity of water that is required for irrigation can usually be obtained from local experts and agronomists or from published tables (Reference 8). It involves several calculation stages:

1. Prediction methods are used to estimate crop water requirements, because of the difficulty of obtaining accurate field measurements.

2. The effective rainfall and groundwater contributions to the crop are subtracted from the crop water requirements to give the net irrigation requirements.

3. Field application and water conveyance efficiency are taken into account to give the gross pumped water requirements.

To illustrate the variation in irrigation water requirements between different crops and different locations, Table 6 gives water requirements for alternative crops in Bangladesh and Thailand.

Month	Bangladesh Apr to Jul: Rice Oct to Apr: Vegetables (m³/day/hectare)	Thailand Jan to Dec: Sugar cane (m³/day/hectare)
January	7	1
February	17	27
March	28	32
April	85	42
May	-	42
June	-	31
July	-	28
August	-	22
September	-	12
October	-	-
November	15	-
December	16	21

Table 6: Typical irrigation water requirements for Bangladesh and Thailand

To find the pumping head

In order to decide what sizes of pump and rotor are required the pumping head must be calculated. The pumping head is the vertical distance that the pump must raise the water.

The total pumping head is the sum of four heights. These are:

1. **Water rest level**
   This is the distance from the ground to the water level when the pump is not running. It is important that this distance is measured during the dry season when the water in the borehole is at its lowest level.

2. **Draw down**
   This is the distance between the water level in the borehole when the pump is not running and the water-level when the pump is running. It will vary with pump speed. The maximum drawdown expected from the borehole can be determined from the test results (see Section Borehole yield on page 44).

3. **Height of the tank**
   This is the height of the top water level of the storage tank above the ground level at the borehole.

4. **Pipework headloss**
   This is the amount of extra lift that the water must be given to overcome the effect of friction between the water and the pipe walls.

Figure 30 illustrates these distances.

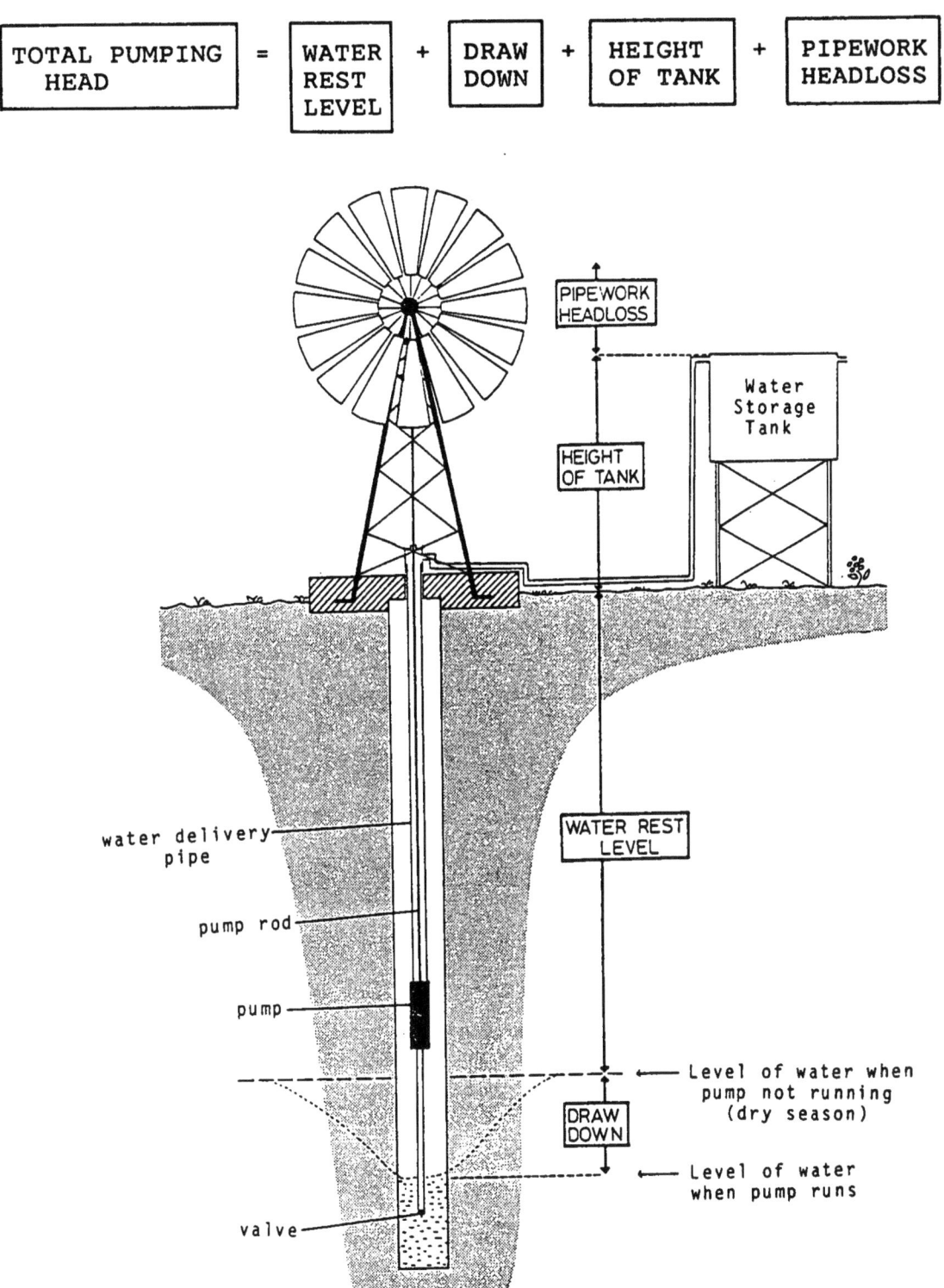

Figure 30: Schematic diagram showing pumping head

The pipework headloss will depend on the following:

1. the length of the pipe
2. the diameter of the pipe
3. the flow-rate of the water through the pipe
4. the roughness of the inside walls of the pipe.

A reasonable estimate of the pipework headloss may be made using Table 7.

The table shows the headloss in metres per 100 m of pipe length, for a range of pipe diameters and water flows. The table should be used to decide on a suitable pipe diameter which gives only a small headloss. The water flow used should be four times the average daily water requirement. (This is because headloss increases rapidly as the flow increases).

Example of use of headloss tables

What diameter of pipe should be used, and what will the headloss be, for a windpump designed to deliver 30 m³ of water per day on average, if the storage tank is 500 m from the pump?

1. Pipe should be designed for four times the average daily water requirement.

    ie. Pipe design flow = 4 x 30
    $$= 120 \text{ m}^3/\text{d}$$

2. Look at Table 7 for a water flow of 120 m³/d, pipe of diameter ≥ 75mm should be used. If 75mm diameter pipe is used the headloss will be 0.35 m per 100 m of pipe.

3. Length of pipe = 500 m (ie. 5 x 100 m)
   Therefore, headloss = 5 x 0.35 = 1.75 m

	PIPE DIAMETER				
mm	25	37	50	75	100
in	1	1	2	3	4
m^3/d					
10	0.50	0.10	0.02	0.00	0.00
20		0.43	0.07	0.01	0.00
30		0.90	0.20	0.02	0.01
40		1.60	0.30	0.04	0.01
50			0.50	0.06	0.02
60			0.70	0.09	0.02
70			1.00	0.12	0.03
80			1.30	0.16	0.04
90			1.60	0.20	0.05
100			2.00	0.25	0.05
110				0.30	0.06
120				0.35	0.08
150				0.60	0.12
200				1.00	0.20
300					0.45

FLOW-RATE

Table 7: Headloss in metres per 100m of pipe length for various flow rates and pipe diameters

Borehole yield

When assessing the water requirement, the potential yield of a borehole in the area should be taken into consideration. This can be found from the test results obtained by the drilling contractor.

When a pump is running in a borehole it has the effect of depressing the water level by an amount which is known as the drawdown (Figure 30). A reasonable estimate of the expected drawdown can also be found from the test results.

It is important that the test results are applicable to the worst month for the borehole, that is when the water level is lowest and the rate of replenishment is at its lowest.

3.3 Sizing the windpump

The windpump must be sized to meet the water demand. In order to do this the amount of energy needed to pump the water must be calculated. A simple way of doing this is to calculate the daily volume-head product in m^4/day for each month. Then the size of both the pump and rotor which can provide this amount of energy can be determined. Finally the size of the storage tank must be determined. This sequence of calculations is illustrated in Figure 31.

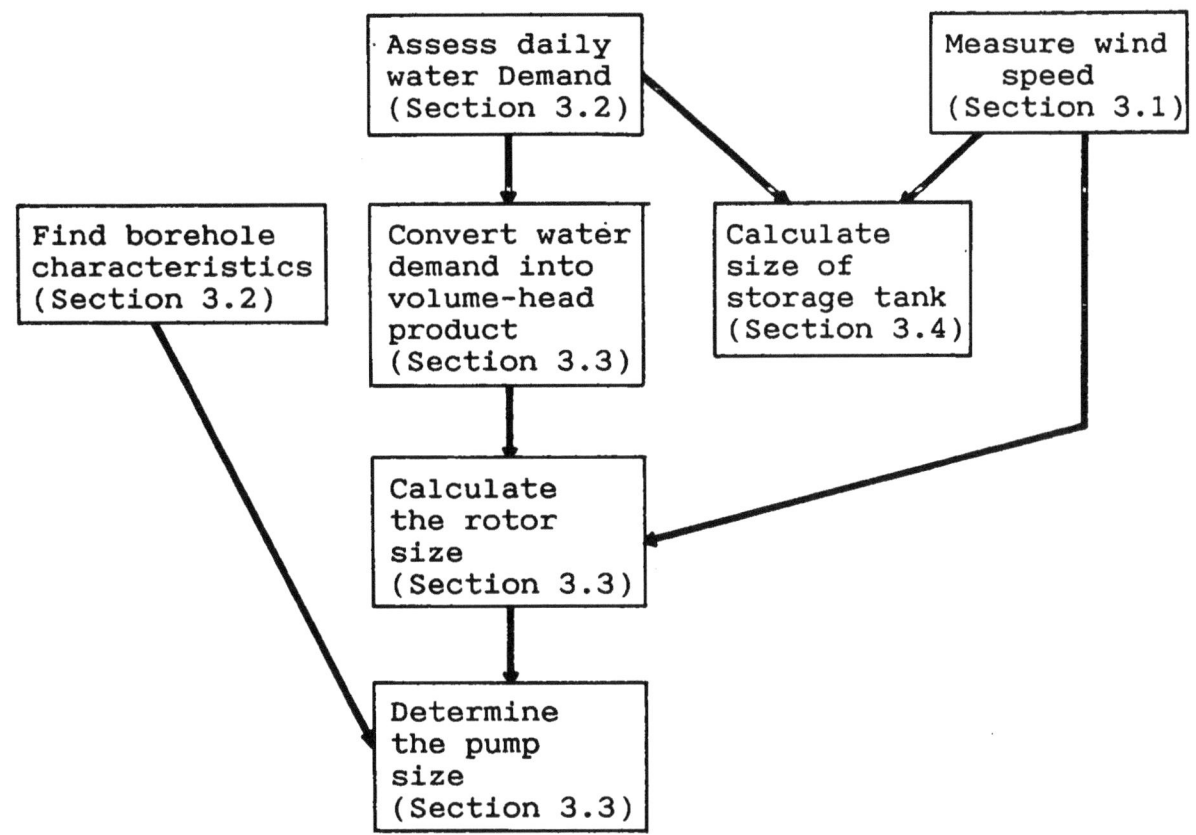

Figure 31: Flow chart outlining the steps necessary to size a windpump

Volume-head product

Once the water requirements have been calculated and the borehole characteristics are known, the effective hydraulic energy needed to provide this amount of water can be determined. The volume-head product is a measure of the hydraulic energy requirements and is simply the product of average daily water requirement and the total pumping head.

| Volume-head product (m^4) | = | Volume of water (m^3) | x | Total head (m) |

Section 3.2 describes how to calculate the total head, which is the sum of the borehole depth, the drawdown, the height of the storage tank and the pipework friction losses.

Example: If the daily water requirement is 31 m^3 per day and the total head is 42 m

$$\text{Volume-head product} = 31 \times 42 = 1302 \ m^4/day$$

Sizing the rotor

The rotor of the windpump must be sized to provide sufficient power to operate the pump when the wind speed is the critical month mean wind speed.

The simplest method for sizing the rotor is to use an empirical equation derived from the Overseas Development Administration Windpump test programme in Kenya. This rule-of-thumb is based on daily measurements on six windpumps over a 12 month period:

$$\boxed{\text{Area of rotor in m}^2} = \boxed{1.14} \times \boxed{\text{Volume-head product in m}^4/\text{day}} \div \boxed{\text{Mean wind speed cubed in m/s}}$$

Continuing the previous example,

The required volume-head product is 1302 m⁴/day.
The mean wind speed is 4 m/s
Hence the required rotor area in m² is

$$\text{Area} = \frac{1.14 \times 1302}{4^3}$$

$$= 23.2 \text{ m}^2$$

This is equivalent to a rotor diameter of 5.4m. In practice the nearest larger size to this could be used, e.g. 6.0 m.

This empirical equation assumes a "typical" performance windpump with "typical" efficiencies and a "typical" tropical wind regime. Despite these assumptions it gives surprisingly accurate first order predictions. It should, however, not be used where the wind regime is highly variable or if good wind speed data is available. In such cases other more complicated methods should be used. They are described fully in Reference 11.

Sizing nomogram

Figure 32 is a rotor-sizing nomogram based on the rule-of-thumb described in the previous section. Calculate the daily volume-head product as described in Section 3.3 and locate this on the horizontal axis. Move vertically upwards until you intersect the correct mean wind speed curve for your location. Now move across to the left and read the rotor diameter off the vertical axis.

The example of the previous section is illustrated on the nomogram.

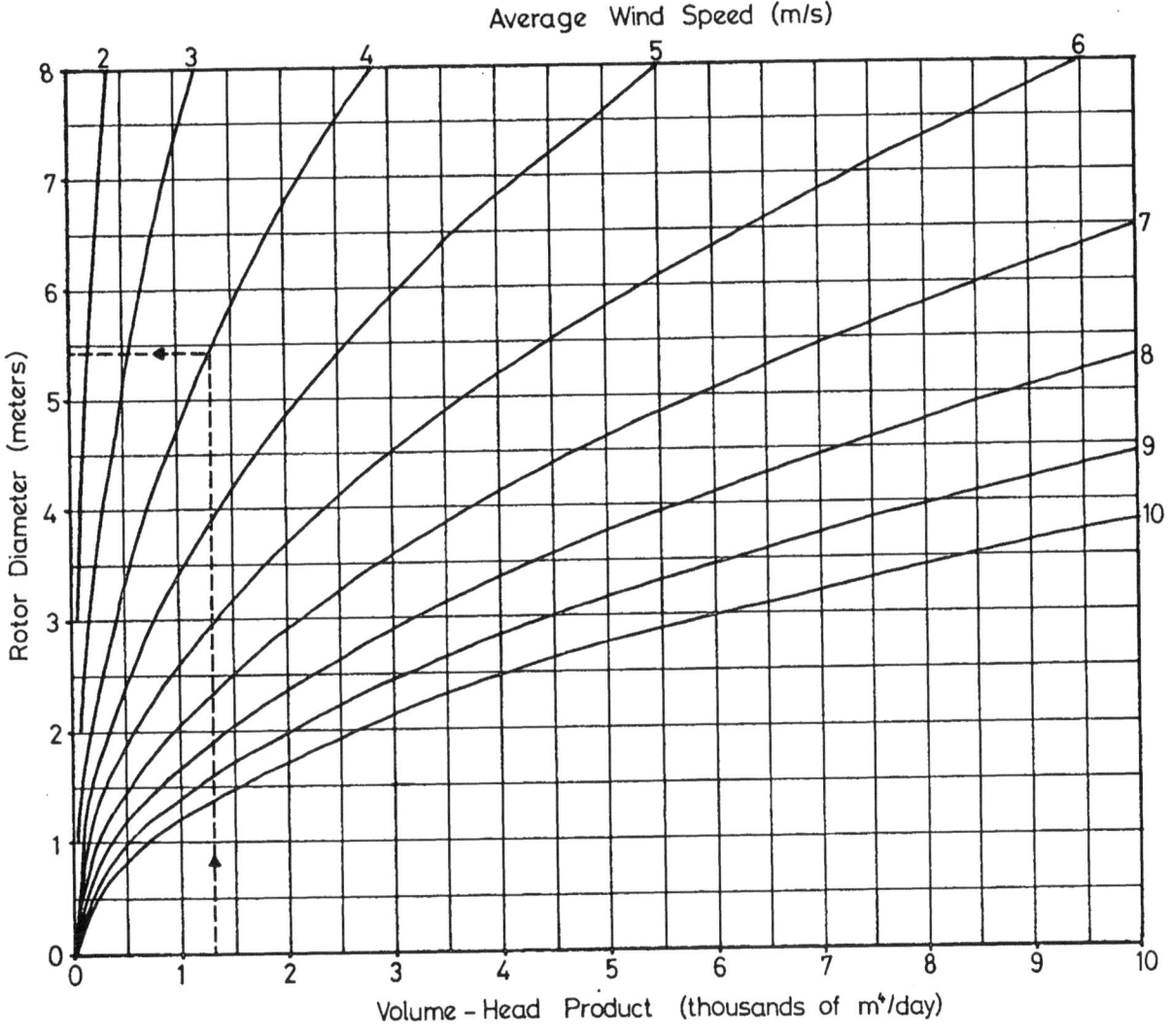

Figure 32: Windpump rotor-sizing nomogram

Sizing the Pump

The pump must be sized to deliver the required amount of water, as efficiently and cheaply as possible. As mentioned in Section 2.3 there are two types of pumps that are commonly used with wind rotors:

1. reciprocating positive displacement pump
2. rotary positive displacement pumps.

The size of the pump may be altered in one of two ways. The pump diameter may be changed, or the pump cylinder length may be changed. In practice the cylinder length of a piston pump, which determines its stroke length, is fixed by the particular design of the transmission, and only the diameter is varied. In general a longer stroke pump is more efficient than a shorter stroke pump of larger diameter which delivers the same volume of water for each stroke. However there are numerous factors which influence the optimum stroke length for a particular diameter, and since the stroke length is fixed by the designer, this will not be considered here.

The pump must be sized so that the starting torque can be matched by the rotor in average winds. If the pump is too large, the rotor will only start to operate at high wind speeds. The pump will then deliver a lot of water at infrequent intervals. If the pump is too small the rotor will start in light winds, but the pump will deliver water at a low flow. It is complicated to calculate the optimum pump size owing to this conflict between making it small enough to start in average winds and large enough to deliver, sufficient water. Obviously the pump size has to be matched to meet the water demand and to be operated by a suitably sized rotor at the wind speed experienced at the site. Fortunately charts are available which summarize pump sizing information. Typical charts are shown in Figure 3.3. These indicate the optimum pump diameter for different pumping heads and average wind speeds (either annual or critical month).

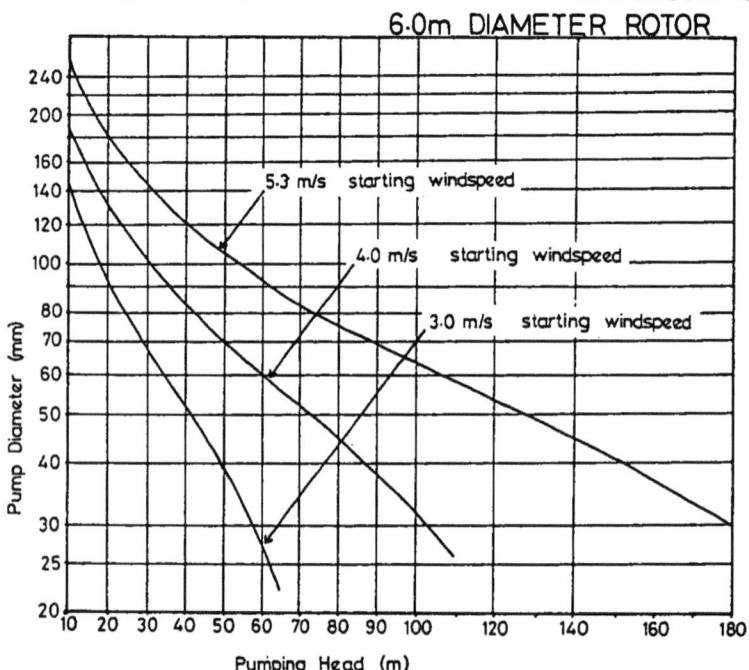

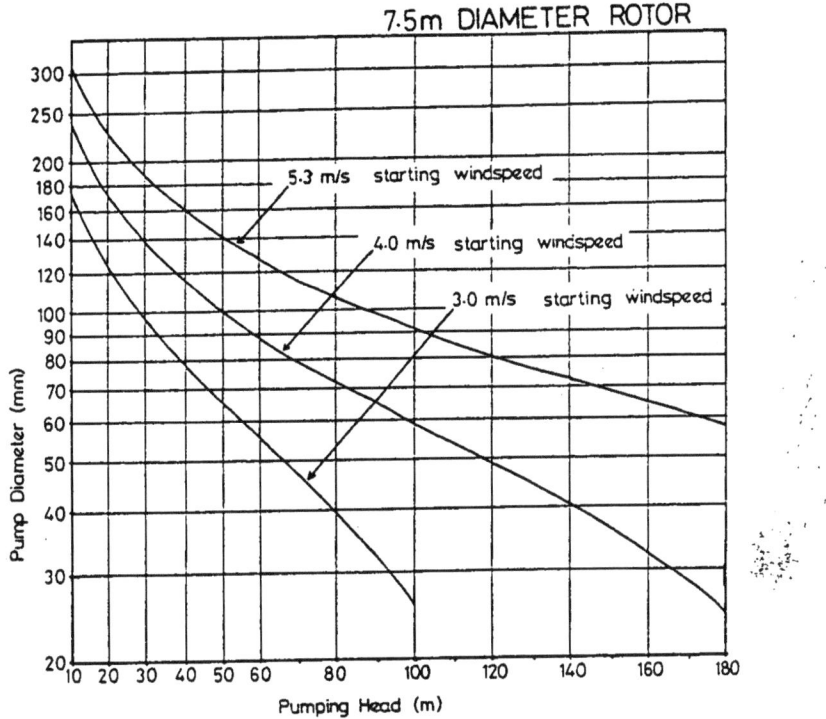

Figure 33: Typical charts for pump sizing by head and average wind speed (suppliers can generally provide sizing advice)

3.4 Storage requirement

Since the wind may not always coincide precisely with the times when water is needed, storage may be required. For irrigation, the economics of water storage should be closely examined since the cost of the tank may not be justified by the additional value of the crops. However, for water supply a tank will certainly be needed. It is important that the storage tank is large enough. Windpumps are sometimes criticized for not supplying sufficient water when the problem is really that the storage tank is too small.

To size the storage tank the following information is needed:

1. the frequency of lull periods each month

2. the duration of the longest lull periods likely to be experienced

3. the daily water requirement.

In this context a lull period is defined as an interval when the windpump does not operate because the wind speed is insufficient to overcome the starting torque of the pump.

In practice, information may not be available about lull periods. In most locations, it is very rare for the wind to be too low to operate the pump more than about three days, so a tank sized for four days would normally be adequate.

The storage capacity is calculated using:

| Storage capacity | = | Daily water requirement | x | Longest lull period | x | Safety factor |

It is advisable to allow a safety margin on the storage tank size, in case longer lull periods do occur, or in case the windpump is out of action for repair or maintenance. A safety factor of two is recommended.

Example

If the daily water requirement is 31 m³/day and the longest lull period is three days,

Storage required = 31 x 3 x 2
= 186 m³

It should be noted that if the size of a storage tank is doubled, the price is less than doubled. In other words the cost per cubic metre of storage decreases as the tank capacity increases. This is because the cost is related to the surface area of the water rather than its volume. Figure 34 illustrates this point for sectional steel tanks. Note that other types of tank may be substantially cheaper.

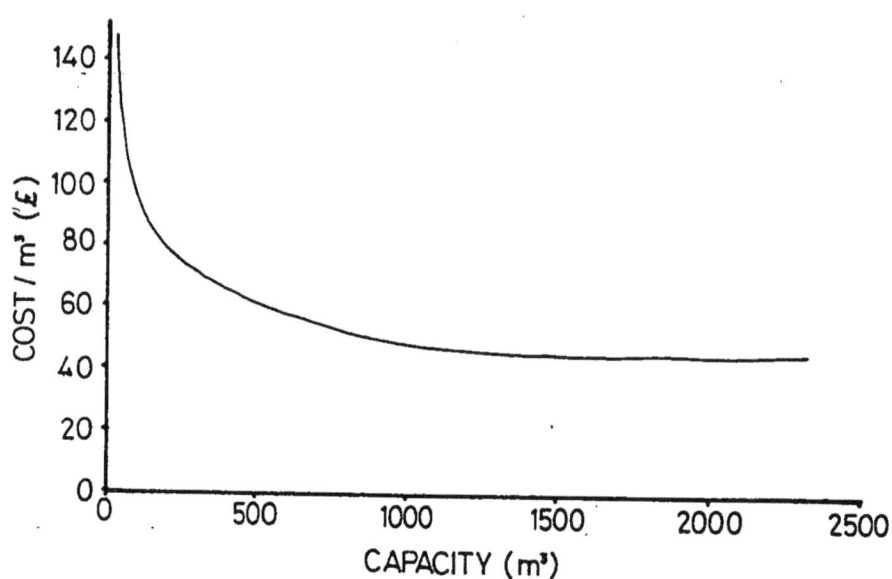

Figure 34: Cost of water storage depends on the volume (based on 1986 UK prices for sectional steel tanks)

Care must be taken using the equation if, at any time during the year, lull periods are longer than the intervening windy periods. If this should be the case, the tank may not completely refill during the windy periods. It will then not be sufficiently full to maintain the water supply during the next lull period.

CHAPTER 4: IS A WINDPUMP THE BEST OPTION?

4.1 The decision route

When considering using a windpump or any other water pumping technology the potential purchaser should follow the decision chart shown in Figure 35.

There are four stages. The first one is to establish the demand for water (refer to Section 3.2). The remaining three stages are discussed in this chapter.

- In Section 4.2 the principal alternatives to windpumps are examined. Graphs and "rules-of-thumb" are given which enable the potential user to determine the size of each alternative. At this stage it may be possible to rule out some alternatives (e.g. if the head is greater than 20 m an animal-driven pump would be unsuitable.

- In Section 4.3 a checklist of social and institutional factors is given. It is most important to ensure that the technology will be acceptable to the users, can be adequately maintained and that there are no institutional constraints that will lead to poor reliability. There is no point choosing what, on paper, appears to be the cheapest option, if it will fail. Again it may be possible to eliminate some options at this stage. For example, if fuel supplies are erratic diesel engines may be rejected.

- Finally, Section 4.4 shows how to cost the alternatives and select the least-cost option. Note that the procedure shown in this section does not determine whether the cheapest option is _economic_, i.e. if the benefits exceed the costs, and for many applications this will be a necessary final stage. There is an important distinction betwen economic and financial project assessment given in this section.

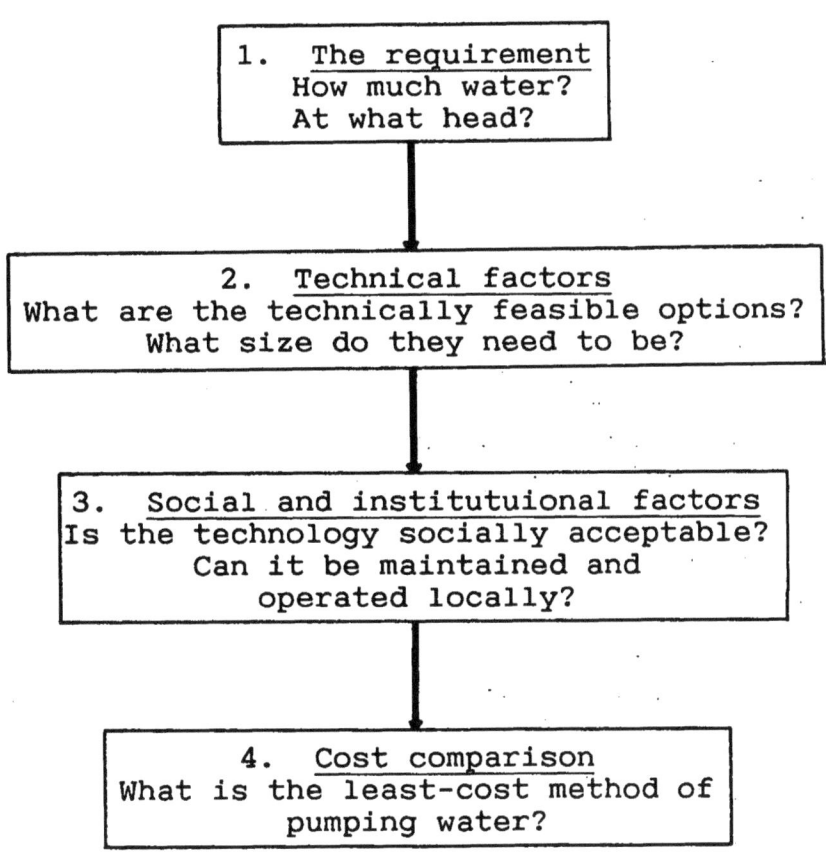

Figure 35: Steps required to choose the most appropriate water pumping technology

4.2 What are the alternatives?

Section 1.4 listed the ways in which energy for water pumping may be supplied and the factors which need to be considered when choosing the most appropriate option. At present the most technically feasible and widespread options for remote applications are:

- grid extension
- diesel, petrol or kerosene engines
- photovoltaic solar pumps
- windpumps
- handpumps
- animal-driven pumps.

This section gives a brief summary of the technical aspects of most of these alternatives to windpumping.

Diesel engines

The internal combustion engine is the world's most common prime mover and it has had more than a century of intensive development. It is a mature technology; however it is sometimes incorrectly applied resulting in uneconomic operation. Diesel engines are often over-sized for small, remote power applications of less than 1 kW. This results in poor part-load performance.

The main characteristics of a diesel pump that should be noted if a diesel engine is being considered are listed below:

1. Power rating
   The power rating required is calculated using:

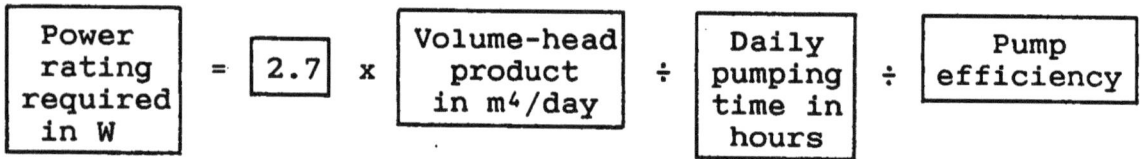

$$\text{Power rating required in W} = 2.7 \times \frac{\text{Volume-head product in m}^4/\text{day}}{\text{Daily pumping time in hours}} \div \text{Pump efficiency}$$

For some applications the required power rating will be less than the smallest commercially available diesel engine. In this case an over-sized engine will have to be used and the engine will have to be either derated or used with a larger pump so that more water is pumped in a shorter time. Derating usually increases the fuel consumption. The derating factor is the ratio of the required power to the power of the engine being used. It is calculated using:

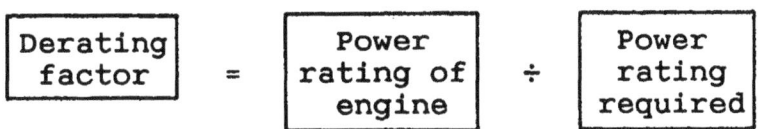

The derating factor must be known in order to estimate the fuel consumption. (Figure 36).

2. Life
Small, lightweight (low cost) diesel engines tend to have short useful lives because they run at high speeds. Wear in machinery is greater at higher speeds. For example, a small 3 kW diesel engine may have a useful life of about 5000 hours between overhauls, whereas a large 50 kW engine will typically achieve over 10,000 hours before sufficient wear has taken place to require a major overhaul.

3. Fuel consumption
Unfortunately it is easy to run an inefficient engine system without realizing it, because any shortfall in performance is compensated by running the engine for a longer period. Fuel consumption is dependent on the derating factor. Typical values are given in Figure 36.

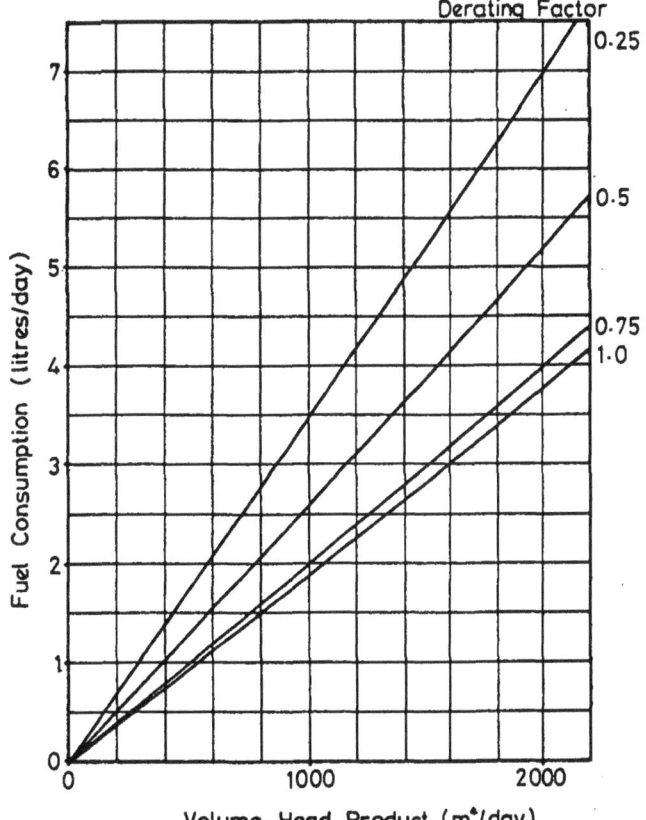

Figure 36: Typical fuel consumptions for small diesel engines

Solar pumps

Solar pump technology is now commercially mature and technically suitable for most water pumping options. There are more than 40 experienced solar pump manufacturers and distributors who have supplied at least 2000 photovoltaic pumping systems. Many are known to be working to the satisfaction of their users. A photovoltaic (PV) pump consists of a series PV modules (termed a PV array), which converts sunlight to electricity. This powers an electric motor-pump unit. For deep boreholes (>10 m) the motor-pump unit is either a submerged motor with a multi-stage centrifugal pump or a surface motor with a submerged rotary pump or piston pump. For low lift applications surface motor-pumps may be used.

Photovoltaic pumps are rated in peak Watts (symbol Wp). This is the power output under peak sunlight conditions. The required rating for a particular application depends on the amount of solar radiation available at the proposed installation site. As with windpumps, a solar pump must be sized to provide sufficient water in the critical month. The critical month is the month in which the ratio of the energy required to the solar energy available is a maximum.

The approximate array size for a solar pump can be calculated using

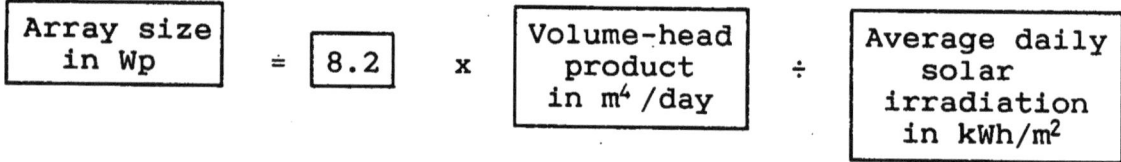

$$\boxed{\text{Array size in Wp}} = \boxed{8.2} \times \boxed{\text{Volume-head product in m}^4\text{/day}} \div \boxed{\text{Average daily solar irradiation in kWh/m}^2}$$

where the volume-head product and the average daily solar irradiation are for the critical month.

For further details on solar pumps refer to References 2 and 3.

Handpumps

Human power is the major source of water lifting in the rural areas of the developing world. There are many traditional water lifting devices such as the Archimedean screw and the shaduf. Handpumps are now widely used, and recently the UNDP/World Bank undertook a major monitoring programme on handpumps. The main technical constraint for a handpump is the amount of power needed to drive it. A healthy adult in a tropical climate can sustain between 40 and 50 watts of power over a 6-7 hour day with short rests (Reference 9). This limits the water output, depending on the total pumped head. Figure 37 shows the number of handpumps that are required to produce different outputs (from Reference 10).

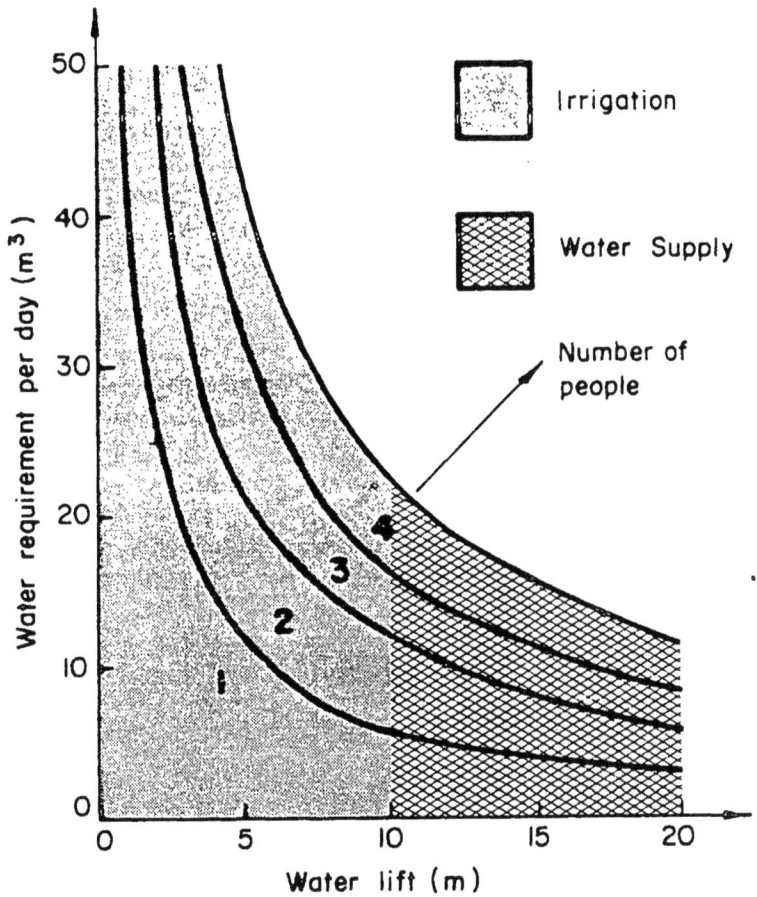

Figure 37: Number of handpumps as a function of water requirement and lift

Animal pumps

Draught animal power is a variable source of energy that depends on a large number of factors, including the species, age, training, physical development and temperament of the animals. Particularly important is whether they are used exclusively for pumping or have other tasks to perform. These variables will affect the power output of the animals and consequently the pump output.

There is little reliable data on the traction power available from animals, but 350 W can be taken as a typical figure. Most animal-driven pumps are low lift (less than 20 m) with efficiencies of 50-70% giving total outputs of 500 - 700 m in an 8 hour day. Figure 38 gives the number of oxen required to provide a range of water outputs at different lifts.

Further details on animal pumps are given in Reference 10.

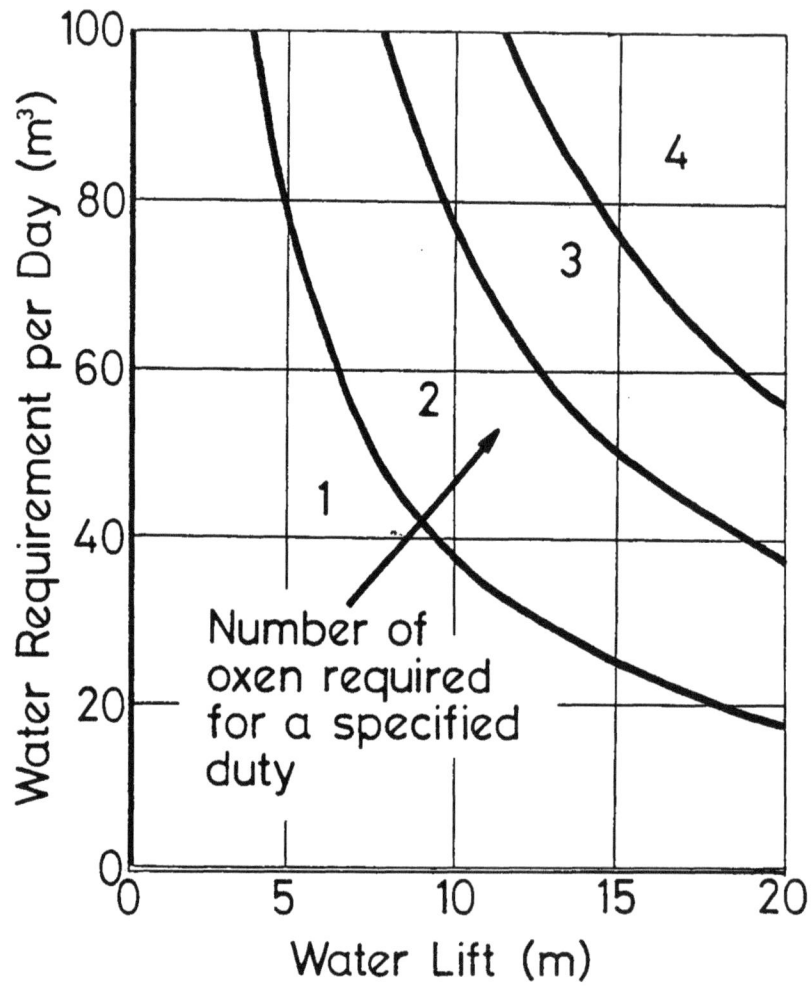

Figure 38: Number of oxen required as a function of water requirement

4.3 Social and institutional factors

There may be objections to a particular method of water pumping for many reasons which are unrelated to the cost. The objections may relate to practical factors such as the availability of fuel or spare parts, or to social acceptability such as the use of animals for drawing water. If a method of water pumping is unacceptable for any such reason it should be eliminated from the list of options before the costs are considered.

Below is a check-list of social and institutional factors some or all of which should be considered for each water pumping method. The list has been divided into three categories; practical factors, social factors and institutional factors.

1. <u>Practical factors</u>

   a) Is the fuel available in sufficient quantity whenever it is needed?
   > For example, is diesel oil always available for diesel pumps or do unacceptable shortages occur? Is there sufficient sunshine all year for solar pumps? Can people provide sufficient energy to meet the water demand using windpumps?

   b) Is the pump manufactured locally, or could it be manufactured locally in the future if demand increased?
   > Local manufacture provides numerous benefits. See Section 2.7.

   c) Are spare parts easily available?
   > There is little point in choosing a technology which on paper is the most cost-effective if spare parts are difficult to obtain, and delivery times are long.

   d) Are all the necessary materials for a particular system readily available?
   > For example, if lifting tackle is necessary, is it available, or if draught animals are needed are there already animals trained to work? Is cement available for making concrete foundations?

   e) If theft is a problem in the locality, some water pumping methods may be considered too vulnerable to being stolen.

   f) If borehole yields are known to be low, to avoid over-pumping high pumping rate devices should not be used. (See Section 3.2).

2. <u>Social factors</u>

   a) Is the proposed scheme wanted by the local people or is some other local improvement (e.g. a road) more important to them? (The feasibility of this will depend on the funding source.)

   b) Is the proposed water supply or irrigation method an improvement over the existing method?

   c) How will the introduction of the pumping technology affect the dependency and power relationships within the community?
   > For example, women collect the water for domestic purposes in many developing countries. If a windpump or solar pump is installed, it may be maintained by men. If an animal-powered pump is installed, men may be responsible for releasing

animals from ploughing duties to enable them to operate the water pump.

d) Will all members of the community, or only the richer members, benefit?
> For example, if a charge is made for water, to cover operation and maintenance, will all members of the community be able to pay the charge?

e) Are the local people able and willing to maintain the water pumping system?

f) Is the water pumping scheme being integrated with other development projects in the community?
> For example, are there also health education programmes, agricultural extension schemes, etc? For animal-powered pumping, are there veterinary services available?

3. Institutional factors

a) If the pumping system has a high capital cost but low recurrent costs, as windpumps and solar pumps have, a large amount of money is needed initially, but very little is needed in subsequent years. For small farmers to purchase such pumping systems, credit facilities must be available to them.

b) If the pumping system has imported components, foreign exchange must be available to pay for them.

c) Are there taxes or import duties which make one pumping option unfavourable, or conversely are there subsidies, or is there easy credit for particular types of pumping system?
> It may be worth lobbying the Government for changes of policies if it is making a pumping method uneconomic by its financial policies.

d) Does the Government have standardized procedures for the implementation of water supply or irrigation projects, or does it have a standardized equipment procurement list?
> If such procedures or lists exist the choice of technologies for water pumping may be restricted.

e) Does the organiszation or community involved have the capacity to implement the project?

4.4 Costing the options

Some water pumping methods will have been eliminated for practical reasons and some for social or institutional reasons. The remaining options should be costed, and the least cost option chosen.

There are a number of different methods of costing. The most common method, life-cycle costing, is recommended and is explained fully in this section. However, first it is important to distinguish between economic and financial assessments.

Economic or financial assessment

There is a distinction between economic and financial assessments. A financial assessment determines the value of a project from the point of view of the purchaser, and as such must include the effects of any taxes, subsidies or interest payments. Interest payments will be made if the purchaser decides to spread the capital cost over several years by taking a loan. An economic assessment determines the value from the point of view of society as a whole and therefore does not include taxes, subsidies or interest payments. An economic assessment attempts to include the value of all benefits, even those which are not readily converted into money, such as improved health. It also assesses the value a resource would have if it were used for a different project. An economic assessment is difficult to do. This section is restricted to the much simpler financial assessment.

Life-cycle costing

Life-cycle costing is a method in which all the costs of the pumping system for its lifetime are added together and expressed as a value in today's money. This value is called the Present Worth of the system.

    Capital cost -   This is the money paid at the beginning for all the components of the pumping system, and their installation. Typically it might include the cost of the windpump, the foundations, drilling the borehole, the pipework, the storage tank, the delivery and labour charges.

    Recurrent costs -   These are the costs of items which are paid for periodically throughout the life of the pumping system. Typically they include the costs of fuel, replacing worn or broken parts, lubricants and labour for regular maintenance and operation duties.

Investment costs - These are the costs associated with investing or borrowing money at the present rate of interest. They take account of the fact that money in the future is worth less than it is now. This is not necessarily because of inflation but simply because it is assumed that money now can be invested and will therefore be worth more in the future. This is explained further in the next paragraph.

If you have £100 you may either spend it or invest it. If the present rate of interest is 10%, and you choose to invest your £100, in one year's time it will be worth £110 or in two years' time it will be worth £110 + 10% of £110 = £121.

So, if you have £110 in one years time, it is the same as having £100 now, or if you have £121 in two years' time it is the same as having £100 now. The Present Worth of spending £110 next year is £100. For example if the cost of diesel oil for a pump set next year is £110, that is equivalent to an expenditure of £100 now. This process of converting future costs into present day costs is called discounting. In this example the discount rate is 10%. The discount rate is the same as the interest rate on the money if it were invested or borrowed. Typically interest rates, and hence discount rates, are 5-20%.

Table 8 gives discount factors to enable present worth to be calculated easily. Choose the appropriate discount rate from the top of the table, and choose the number of years over which you wish to discount from the side of the table. Where the discount rate column and the number of years row intersect is the discount factor. For example, to find the present worth of a cost of £250 in four years' time, at a discount rate of 10%, the discount factor is 0.68.

$$\text{Present Worth} = £250 \times 0.68$$
$$= £170$$

If a cost has to be paid every year over the life of the pump, it is tedious to use Table 8 to calculate the present worth. Table 9 has been included to make such calculations easier. For example, if the annual maintenance costs for a pump are £350 for 12 years at a discount rate of 8%:

$$\text{the Present Worth} = £350 \times 7.54$$
$$= £2638$$

		DISCOUNT RATE %								
		2	4	6	8	10	12	14	16	18	20
NUMBER OF YEARS	1	0.98	0.96	0.94	0.93	0.91	0.89	0.88	0.86	0.85	0.83
	2	0.96	0.92	0.89	0.86	0.83	0.80	0.77	0.74	0.72	0.69
	3	0.94	0.89	0.84	0.79	0.75	0.71	0.67	0.64	0.61	0.58
	4	0.92	0.85	0.79	0.74	0.68	0.64	0.59	0.55	0.52	0.48
	5	0.91	0.82	0.75	0.68	0.62	0.57	0.52	0.48	0.44	0.40
	6	0.89	0.79	0.70	0.63	0.56	0.51	0.46	0.41	0.37	0.33
	7	0.87	0.76	0.67	0.58	0.51	0.45	0.40	0.35	0.31	0.28
	8	0.85	0.73	0.63	0.54	0.47	0.40	0.35	0.31	0.27	0.23
	9	0.84	0.70	0.59	0.50	0.42	0.36	0.31	0.26	0.23	0.19
	10	0.82	0.68	0.56	0.46	0.39	0.32	0.27	0.23	0.19	0.16
	15	0.74	0.56	0.42	0.32	0.24	0.18	0.14	0.11	0.08	0.06
	20	0.67	0.46	0.31	0.21	0.15	0.10	0.07	0.05	0.04	0.03

Table 8: Discount factors for various discount rates and number of years (zero inflation)

		DISCOUNT RATE %								
		2	4	6	8	10	12	14	16	18	20
NUMBER OF YEARS	1	0.98	0.96	0.94	0.93	0.91	0.89	0.88	0.86	0.85	0.83
	2	1.94	1.88	1.83	1.79	1.74	1.69	1.65	1.60	1.57	1.52
	3	2.88	2.77	2.67	2.58	2.49	2.40	2.32	2.24	2.18	2.10
	4	3.80	3.62	3.46	3.32	3.17	3.04	2.91	2.79	2.70	2.58
	5	4.71	4.44	4.21	4.00	3.79	3.61	3.43	3.27	3.14	2.98
	6	5.60	5.23	4.91	4.63	4.35	4.12	3.89	3.68	3.51	3.31
	7	6.47	5.99	5.58	5.21	4.86	4.57	4.29	4.03	3.82	3.59
	8	7.32	6.72	6.21	5.75	5.33	4.97	4.64	4.34	4.09	3.82
	9	8.16	7.42	6.80	6.25	5.75	5.33	4.95	4.60	4.32	4.01
	10	8.98	8.10	7.36	6.71	6.14	5.65	5.22	4.83	4.51	4.17
	15	12.8	11.1	9.7	8.6	7.6	6.8	6.2	5.6	5.1	4.6
	20	16.3	13.6	11.5	9.8	8.5	7.5	6.6	5.9	5.4	4.8

Table 9: Discount factors for recurrent costs which have to be paid annually over a number of years, for various discount rates (zero inflation)

It is usual to carry out economic assessments in real terms. The interest rates and discount rates used should be relative to general inflation. If both discount and inflation rates are expressed in the same way (i.e. both excluding general inflation or both including general inflation) the resulting Present Worth will be the same.

A more detailed example, in which all the future recurrent costs are discounted to the present and added to the capital cost, is described in the next section.

The steps needed are summarized in the flow chart in Figure 39.

Example financial assessment

The following example illustrates the steps shown in the flow chart on the previous page. In this example, solar, hand and animal pumps have been eliminated for a variety of technical, practical, social and institutional reasons. The final choice is between a diesel-powered pump and a windpump.

The data needed to compare the costs are given below:

Daily Water Demand	= 31 m^3
Total Pumping Head	= 42 m
Period of Financial Assessment	= 10 years
Discount Rate (excluding general inflation)	= 10%
Inflation Rate (excluding general inflation)	= 0%
Mean Annual Wind Speed	= 4 m/s
Longest Lull Period	= 5 days
Price of Diesel Oil	= £0.30 per litre

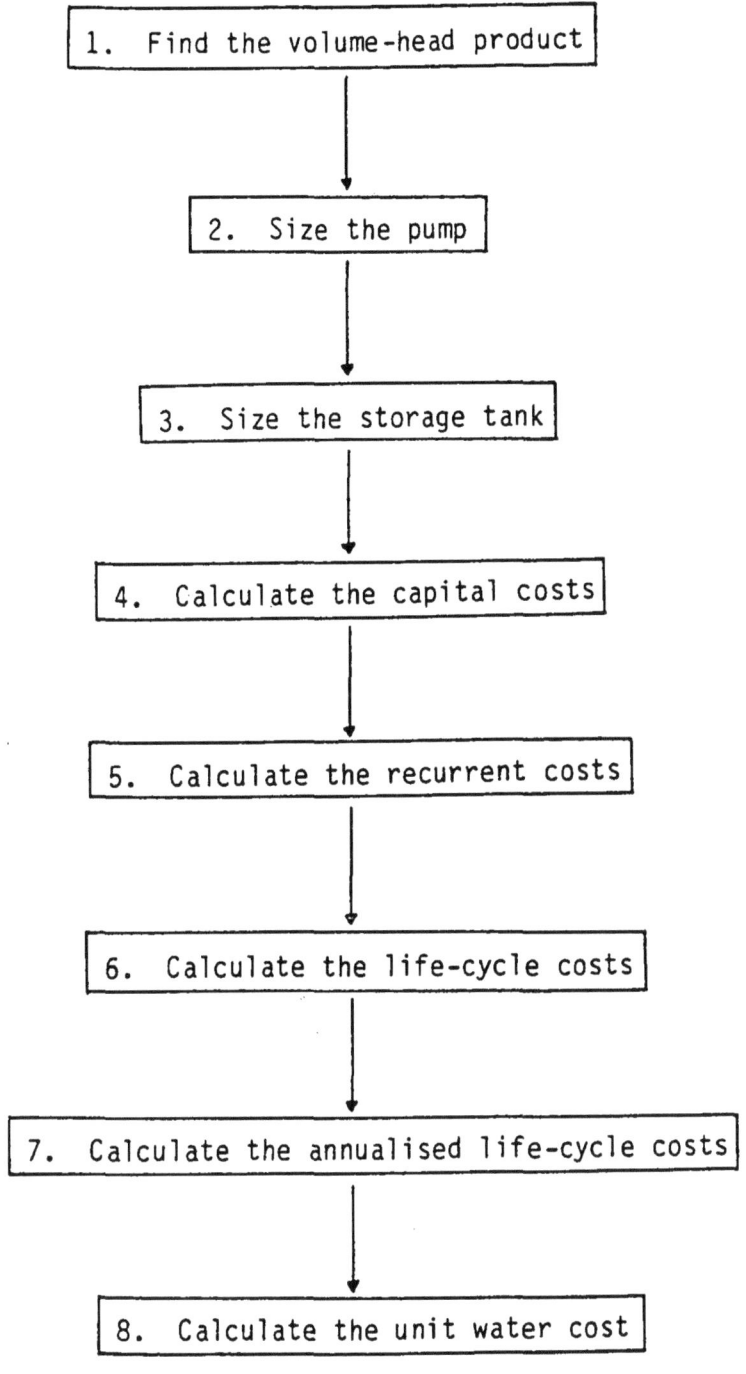

Figure 39: Flow chart showing the steps to be taken in financial assessment

Step 1: Find Volume-head product

　　　　Volume-head product = 31 m³ x 42 m
　　　　　　　　　　　　　　= 1302 m⁴/day

Step 2: Size the pump

WINDPUMP

From nomograms (pages 47 and 49) using 1302 m⁴/day at 4 m/s wind speed,

　　Rotor diameter = 6 m
　　Pump diameter = 75 mm

DIESEL PUMP

From nomograms (page 55) using 1302 m⁴/day and assuming 7.5 % overall efficiency,
　　Diesel engine size = 5 kW
　　Fuel consumption = 4.6 litres/day

Step 3: Size the storage tank

WINDPUMP

Using equation from page 50 with safety factor of 2.

　　Tank size = 5 x 31 x 2
　　　　　　= 310 m³

DIESEL PUMP

Allow 3 days storage to cover maintenance and repair periods. Use a safety factor of 2.
　　Tank size = 3 x 31 x 2
　　　　　　= 186 m³

Step 4: Calculate the capital costs

WINDPUMP

6m Windpump	= £ 7000
310 m³ concrete tank at £15/m	= £ 4650
Borehole construction 35 m at £100/m	= £ 3500
Delivery pipe 100m at £2/m	= £ 200
Foundations	= £ 100
Delivery and installation	= £ 1000
Total	£16450

DIESEL PUMP

5kW Diesel pump	= £4000
186 m³ concrete tank at £25/m	= £4650
Borehole construction 35 m at £100/m	= £3500
Delivery pipe 100m at £2/m	= £ 200
Foundations	= £ 100
Delivery and installation	= £ 300
Total	£12750

Note: The cost per cubic metre of storage tank is lower for a larger tank.

Step 5   Calculate the recurrent costs

WINDPUMP	DIESEL PUMP
Fuel = nil	Fuel (4.6 litres/day at £0.30 per litre for 365 days) = £ 504
Operation and maintenance = £ 100/yr	Operation and maintenance = £ 400
Total annual recurrent cost = £ 100	Total annual recurrent cost = £ 904

Discount at 10% for 10 years using Table 9

WINDPUMP:
Present worth = £100 x 6.14
            = £614

DIESEL PUMP:
Present worth = £904 x 6.14
            = £5549

Delivery pipe is replaced after 5 years at a cost of £200. If this is discounted

WINDPUMP:
Present worth = £200 x 0.62
            = £124

DIESEL PUMP:
Present worth = £200 x 0.62
            = £124

TOTAL Present worth of recurrent costs

WINDPUMP: = £514 + £124 = £740

DIESEL PUMP: = £5549 + £124 = £5670

Step 6   Calculate the life-cycle cost

WINDPUMP
Capital cost   = £16450
Recurrent cost = £ 740
Life-cycle cost = £17190

DIESEL PUMP
Capital cost   = £12750
Recurrent cost = £ 5670
Life-cycle cost = £18420

Note: The windpump, over ten years, is 7% cheaper than the diesel pump. Over a longer time, it would probably be much more than 7% cheaper because the diesel engine will wear out and need to be completely replaced before the windpump will.

The comparison could stop here. However, sometimes it is useful to compare unit water costs (i.e. the cost per cubic metre of pumped water). If this is required, steps 7 and 8 should be followed.

Step 7: Calculate the annualized life-cycle cost

In this step the total life-cycle cost is spread over the period of the financial assessment. This is done by the reverse of discounting. The life-cycle cost is divided by the appropriate factor from Table 9.

WINDPUMP

Discount at 10% for
10 years, using Table 9

Annualized life-cycle

 cost = £17190 ÷ 6.14

   = £2800

DIESEL PUMP

Discount at 10% for
10 years, using Table 9

Annualized life-cycle

 cost = £18420 ÷ 6.14

   = £3000

Step 8: Calculate the unit water cost

 Total annual water demand = 31 m^3 × 365 days

          = 11315 m^3

WINDPUMP

Unit water
cost   = £2800 ÷ 11315

    = 24.7 p/m^3

DIESEL PUMP

Unit water
cost   = £2150 ÷ 11315

    = 26.5 p/m^3

Note: It may be useful to know the unit water cost, if a water rate is to be paid by each member of the community.

CHAPTER 5: PROCUREMENT, INSTALLATION AND OPERATION

5.1 Specifying and procuring

Five stages for the specification and procurement of windpumps are recommended, as described briefly below:

1. **Assess windpumping viability and estimate the costs**
   Before contacting suppliers, an initial appraisal of windpumping should be made in accordance with the guidelines given in Chapters 1, 3 and 4.

2. **Prepare tender documents**

3. **Issue a call for tenders**
   Send letters to suppliers, especially local ones, with a brief description of the required systems. Ask interested suppliers to reply requesting tender documents.

4. **Preliminary evaluation**
   Each tender should be checked to ensure that:
   - the windpump offered is complete and includes spare parts, tools and installation and operating instructions
   - the windpump offered can be delivered within the time specified
   - an appropriate warranty can be provided.

5. **Detailed assessment**
   A detailed assessment of each tender should be made, with respect to four aspects. Approximately equal importance should be ascribed to each of the four items.

   <u>Compliance with specification</u> Check the windpump output at the mean annual wind speed, and check its overall efficiency at this speed. Check the wind speeds at which the windpump starts to pump water due to there being sufficient torque and stops pumping water due to it furling. Is this operating wind speed range adequate?

   > **Note:** Beware of some suppliers who quote instantaneous flow rates in l/s or m^3/h at high windspeeds. A pump which operates well at a high wind speed will probably not operate at all at a low wind speed.

   <u>System design</u> The suitability of the windpump should be assessed taking into account operation and maintenance requirements, complexity of the transmission and other parts, safety features, etc.

The life of the equipment should be assessed, by examining the parts liable to wear and tear e.g. bearings, pump rods, fixing of rotor blades, water-tight seals, etc.

The content of the information supplied to support the tender should be assessed with respect to the provision of general assembly drawings and performance information. The performance information should include wind speed versus power and torque curves for the rotor, and head versus flow for a range of speeds for the pump. Rotor and pump efficiency curves should also be supplied.

<u>Capital cost</u> Capital costs should be compared, allowing for any deviations from the specification proposed by the tenderer.

<u>Overall credibility</u> The experience and resources of the tenderer relevant to windpump technology in developing countries should be assessed. The tenderer should be able to provide a repair and spare parts service within a reasonable time if problems with the windpump arise. A warranty should be provided. It is recommended that the warranty should cover at least five years.

5.2 Installation

The site

The site for the windpump should have been selected during the wind speed measurements. To re-cap, the site should be located:

1. to receive maximum average wind

2. to catch prevailing winds, on a nearby hill, etc.

3. to avoid turbulence i.e. far from buildings and tall trees, on as smooth ground as possible, and on a tower sufficiently high that the rotor is in a uniform air stream. Note: If the upper part of the rotor consistently receives greater winds than the lower part, the uneven loading is likely to cause early rotor failure

4. where there is adequate ground water.

Receipt of windpump

When the windpump is received, carefully check all parts for damage in transit. Check that you have received all necessary documentation including assembly and installation instructions, repair and maintenance instructions, a warranty agreement and a list of parts and spares.

Store all parts in a secure place until you are ready to install the windpump.

Borehole construction

It is not the intention of this handbook to give detailed information about how to construct a borehole. Such information exists in numerous other publications. However there are a number of points which must be taken into consideration during borehole construction. These are listed below:

1. The borehole must be deep enough for the pump suction pipe to be always submerged, even in the dry season with the pump operating at maximum speed (and hence with maximum drawdown).

2. The borehole screening must allow the borehole to recharge as fast as water is pumped out, at all times of year.

3. The gravel filling in the annulus between the drilled hole and the screening must be sufficient to prevent sand and other fines from silting up the borehole, clogging the pump and causing excessive wear of pump parts.

4. The screen must be made of a suitable material for the prevailing conditions. Steel or galvanized iron is likely to corrode rapidly. Stainless steel will last a long time but it is expensive. Plastics are cheap and lightweight but may deform under load, especially in deep wells. Burnt-clay pipe screens are brittle.

After drilling the borehole, water samples should be taken and tested for pH and salinity. The chemistry teacher at the local secondary school should be able to help with this. If the pH or salinity is high the water may be unsuitable for drinking or irrigation. It may also cause rapid corrosion of some pump parts.

Windpump foundations

The windpump manufacturer should provide instructions about the type and dimensions of the foundations, and the details of the fastening of the tower to the foundations. The following points should be taken into consideration:

1. It is essential that the pump or borehole is in the centre of the tower. Therefore care must be taken to build the foundations in the correct location.

2. Some windpumps can be laid on their side for repair and maintenance. Sufficient space should be allowed around the site for this.

3. The foundations must be massive enough to withstand the maximum forces on the windpump structure in a storm.

4. The mouth of the borehole should be concreted at the same time as the foundations for the windpump legs are being built. The concrete should slope away from the mouth of the borehole to prevent surface water from flowing in and contaminating it.

5. The concrete must be kept damp at all times whilst it is curing (i.e. for about a week) otherwise cracks will develop and weaken it.

Erecting the tower and assembling the rotor

The tower may be built vertically on top of the concrete base or it may be assembled horizontally and then lifted. Follow the manufacturer's instructions at all times, but before starting READ THE SAFETY INSTRUCTIONS in the manufacturer's literature and on page 75 of this Handbook.

It is much easier if the rotor can be assembled close to the ground and then lifted into position.

Building the storage tank and delivery pipe

The storage tank should be as close as is reasonably possible to the windpump, to minimize the length of delivery pipe. The advantages and disadvantages of various construction methods are given in Table 10. The most important point to note is that the tank must be watertight. A tiny leak of one drip per second will waste five litres of water per day.

Construction method	Advantages	Disadvantages
Prefabricated plastic	Easy Leak-tight Corrosion resistant	Small sizes only Expensive Will slowly deteriorate in sunlight Prone to rodent and termite attack
Prefabricated galvanized iron	Easy Leak-tight	Small sizes only Will corrode eventually as galvanizing wears through
Site-assembled steel plate	All sizes possible Robust Leak-tight if correctly assembled	Expensive Requires painting or lining to prevent corrosion
Site-assembled GRP plate	All sizes possible Leak-tight if correctly assembled Corrosion resistant	Expensive Will slowly deteriorate in sunlight
Stone, brick or concrete	All sizes possible Cheap Materials locally available	Large tanks need walls/buttresses to support weight of water Require waterproof cement on inside Liable to leak

Table 10: Advantages and disadvantage of various construction methods for storage tanks

A delivery pipe of adequate diameter (see Section 3.2) must be laid to the tank. This must <u>not</u> have a gate-valve on it, or a float valve on the tank inlet, because if a valve were shut such that the windpump was pumping against the closed valve, the pump would be seriously damaged. The tank should be provided with an overflow pipe which discharges into a nearby stream. The delivery pipe may be plastic, galvanized iron, cast iron or any other material. No material is perfect; iron corrodes, plastics are eaten by rodents. Make sure that there are no leaks in the delivery pipe.

Fences, etc

Build a fence round the windpump to prevent people or animals being injured by the moving parts.

Protect the delivery pipe from accidental damage by vehicles, people or animals. Note that plastic pipes are easily damaged by people walking on them.

5.3 Maintenance and repair

The efficiency, economy, safety and length of service of any piece of equipment will be proportional to the standard of maintenance. Windpumps require only a little maintenance, but that little maintenance is important. You should follow the manufacturer's instructions for maintenance, but bear in mind the following:

1. The rotor, tail and tower should have been designed and manufactured to withstand severe winds and storms. However, they will need to be checked from time to time for cracks, etc., especially at welds, around bolts and at sharp changes in sections.

2. If the windpump is correctly installed and well-maintained it should operate fairly quietly. If you hear any unusual noise you should investigate.

3. Moving parts must be lubricated from time to time to prevent them from wearing. It is recommended that you grease the transmission and any pump rod guides at least once every six months.

4. If any parts are worn, replace them.

5. Repaint parts of the windpump from time to time to prevent corrosion.

6. Inspect the delivery pipe and storage tank periodically for leaks.

If a windpump is well-maintained it should not need to be repaired more than occasionally. Pump seals and valves may need replacement from time to time. Those and other simple replacements should be easy to do using spare parts and tools supplied by the manufacturer. If major repairs are necessary it is advisable to contact the manufacturer (the advantage of purchasing from a local manufacturer becomes evident here). Remember that your warranty may cover you for some types of repairs.

It is advisable to check the water level in the borehole at least once a week during the first year of operation, to ensure that the aquifer is recharging the borehole sufficiently.

5.4 Safety

It is advisable to take the safety precautions listed below to prevent people from being injured.

1. **Never** work on a windpump **alone.**

2. Always have at least **two people** when maintaining or working on the windpump.

3. Keep **fingers** and **toes well clear** of any moving parts.

4. Try **not to stand underneath** the windpump if someone is working up it.

5. **Be aware** of the behaviour of the blades when climbing the tower.

6. **Do not let children climb** on the windpump.

7. **Wear a hard hat** if possible.

It is advisable to take the safety precautions listed below to prevent damage to the machinery:

1. **Build a fence** round the windpump

2. **Protect the delivery pipe**

3. **Check** that the **windpump furling mechanism works** periodically.

4. **Do not put a gate valve** in the delivery pipeline, **or a float valve** on the storage tank (see Section 5.2).

BIBLIOGRAPHY

1. World Meteorological Technical Note on Wind Energy. (Map 1981).

2. Water-Pumping Devices, A handbook for users and choosers by Peter Fraenkel. Publisher: I.T. Publications (1986).

3. Solar Water Pumping. A Handbook by Jeff Kenna and Bill Gillett. Publisher: I.T. Publications (1985).

4. Water Current Turbines, A Fieldworker's Guide by Peter Garman. Publisher: I.T. Publications (1986).

5. Small Business in the Third World - Guidelines for practical assistance by Malcolm Harper. Publisher: I.T. Publications (1984).

6. A Siting Handbook for Small Wind Energy Conversion Systems. Publisher: WindBooks (1980).

7. Wind Technology Assessment Study. Report by I.T. Power Ltd. for UNDP Project GLO/80/003 executed by the World Bank. (1983).

8. Crop Water Requirements by Doorenbos and Pruitt. Irrigation and Drainage Paper no. 24 (1977). FAO, Rome.

9. Water Lifting/Water Pumping Study - Chad. Report by I.T. Power Ltd for USAID. (1986).

10. Human and Animal-Powered Water-Lifting Devices, A state-of-the-art survey by W.K. Kennedy and T.A. Rogers. Publisher: I.T. Publications (1985).

11. Introduction to Wind Energy by E.H. Lysen. Publisher: SWD (1982).

12. Wind Systems Life-Cycle Cost Analysis by J.M. Sherman, M.S. Gresham and D.L. Fergason. Publisher: WindBooks (1983).

GLOSSARY

Aerofoil - a shape which creates a large lift force relative to drag force when wind passes across it (e.g. an aeroplane wing or a windpump blade).

Anemometer - an instrument for measuring wind speed.

Annualized life-cycle cost - total life-time costs of a pumping system expressed as a sum of annual payments.

Capital cost - a cost which is paid only once, at the time of the initial purchase.

Centrifugal pump - a pump which raises water by the action of centrifugal force on the water as it passes radially through a rotating impeller.

Critical month - the month in which the water demand is greatest in relation to the energy (wind or solar) available for pumping.

Darrieus rotor - a vertical axis two-bladed rotor of low solidity.

Discount rate - the annual rate by which future costs are reduced to convert them into equivalent present day costs.

Drag - a force produced by the wind on objects in its path, which is in the same direction as the wind.

Draw down - the difference in water level in a borehole when the pump is running and when it is not.

Field capacity - the maximum amount of water that soil can hold before it becomes saturated.

Furl - the rotation of a wind rotor to be parallel with the wind in stormy weather, or for maintenance.

Head (pumping) - the total distance that a pump raises water.

Headloss - the head destroyed by friction in pipelines, etc.

Kinetic energy - energy due to motion.

Life-cycle costs - the lifetime costs of a pumping system expressed in present day money.

Lift (pump) - see head (pumping).

Lift (wind) - a force produced by the wind on objects in its path, which is perpendicular to the direction of the wind.

Lull period - an interval when the wind speed is insufficient to operate the windpump.

Panamone - vertical axis wind rotor which utilizes the drag force of the wind.

Performance coefficient - the fraction of wind energy passing through a rotor disc which is converted into shaft power.

Photovoltaic pump - a pump which is powered by photovoltaic cells. These convert sunlight directly into electricity.

Piston pump - see reciprocating positive displacement pump.

Pitch - the angle at which wind rotor blades are set to the wind.

Positive displacement pump - a pump which lifts water by displacing it upwards.

Progressive cavity pump - see rotary positive displacement pump.

Present worth - the value of a future cost or benefit expressed in present-day money.

Prime mover - the power source for a pumping system.

Reciprocating positive displacement pump - a positive displacement pump in which the water lifting components of the pump move up and down.

Recurrent cost - a cost which is paid periodically throughout the life of the pumping system.

Regression analysis - a mathematical technique for data comparison.

Rotary positive displacement pump - a positive displacement pump in which the water-lifting components of the pump rotate.

Rotor (pump) - the element of a pump which rotates and lifts water.

Rotor (wind) - the part of a windmill which is caused to rotate by the wind blowing at it.

Savonius rotor - vertical-axis, high-solidity wind rotor which chiefly utilizes the drag force of the wind.

Solidity - the fraction of the swept area of a wind rotor which is filled with blade.

Stator - the stationary part of a rotary positive displacement pump, within which the pump rotor rotates.

Tip-speed ratio - the ratio of the speed of a wind rotor's blade tips to the speed of the wind.

Torque - the turning force produced by a wind rotor, or required by a pump.

Transmission - those components of a windpump which convert the rotation of the wind rotor into a suitable movement to drive the pump.

Turbulence - unsteady, circulating air flow.

Unit water cost - the cost of providing one cubic metre (or one litre) of water.

Volume-head product - the product of the quantity of water required and the head through which the water must be lifted, expressed in m^4.

Water rest level - the water level in a borehole or well when the pump is not running.